British Migration

T0199715

Around 5.6 million British nationals live outside the United Kingdom: the equivalent of one in every ten Britons. However, social science research, as well as public interest, has tended to focus more on the numbers of migrants entering the UK, rather than those leaving.

This book provides an important counterbalance, drawing on the latest empirical research and theoretical developments to offer a fascinating account of the lives, experiences and identities of British migrants living in a wide range of geographic locations across Europe, Asia, Africa and Australasia. This collection asks: What is the shape and significance of contemporary British migration? Who are today's British migrants and how might we understand their everyday lives? Contributions uncover important questions in the context of global and national debates about the nature of citizenships, the 'Brexit' vote, deliberations surrounding mobility and freedom of movement, as well as national, racial and ethnic boundaries.

This book challenges conventional wisdoms about migration and enables new understandings about British migrants, their relations to historical privileges, international relations and sense of national identity. It will be valuable core reading to researchers and students across disciplines such as Geography, Sociology, Politics and International Relations.

Pauline Leonard is Professor of Sociology at the University of Southampton, UK.

Katie Walsh is Senior Lecturer in Human Geography at the University of Sussex, UK.

Routledge Studies in Human Geography

This series provides a forum for innovative, vibrant, and critical debate within Human Geography. Titles will reflect the wealth of research which is taking place in this diverse and ever-expanding field. Contributions will be drawn from the main sub-disciplines and from innovative areas of work which have no particular sub-disciplinary allegiances.

Geographical Gerontology
Edited by Mark Skinner, Gavin Andrews, and Malcolm Cutchin

New Geographies of the Globalized World
Edited by Marcin Wojciech Solarz

Creative Placemaking
Research, Theory and Practice
Edited by Cara Courage and Anita McKeown

Living with the Sea
Knowledge, Awareness and Action
Edited by Mike Brown and Kimberley Peters

Time Geography in the Global Context
An Anthology
Edited by Kajsa Ellegård

Consolationscapes in the Face of Loss
Grief and Consolation in Space and Time
Edited by Christoph Jedan, Avril Maddrell and Eric Venbrux

The Crisis of Global Youth Unemployment
Edited by Tamar Mayer, Sujata Moorti and Jamie K. McCallum

Thinking Time Geography
Concepts, Methods and Applications
Kajsa Ellegård

British Migration
Globalisation, Transnational Identities and Multiculturalism
Edited by Pauline Leonard and Katie Walsh

For more information about this series, please visit: www.routledge.com/ Routledge-Studies-in-Human-Geography/book-series/SE0514

British Migration

Privilege, Diversity and Vulnerability

Edited by
Pauline Leonard and Katie Walsh

LONDON AND NEW YORK

First published 2019
by Routledge
2 Park Square, Milton Park, Abingdon, Oxon OX14 4RN

and by Routledge
52 Vanderbilt Avenue, New York, NY 10017

First issued in paperback 2020

Routledge is an imprint of the Taylor & Francis Group, an informa business

British Library Cataloguing in Publication Data
A catalogue record for this book is available from the British Library

Library of Congress Cataloging-in-Publication Data
A catalog record has been requested for this book

ISBN 13: 978-0-367-58269-2 (pbk)
ISBN 13: 978-1-138-69033-2 (hbk)

Typeset in Times New Roman
by Taylor & Francis Books

Contents

9 Size matters: British women's embodied experiences of size in Singapore 146
JENNY LLOYD

10 Resilience and social support among ageing British migrants in Spain 164
KELLY HALL

11 Returning at retirement: British migrants coming 'home' in later life 182
KATIE WALSH

Index 199

Contributors

Gillian Abel is a doctoral graduate from The University of Western Australia whose thesis focused on contemporary British migration to Western Australia. Her broader research interests include lifestyle migration and migration hierarchies. She is currently employed in an unrelated field.

Michaela Benson is Reader at Goldsmiths, University of London. She is internationally renowned for her contributions to the sociology of migration and is currently a research leader for the project BrExpats: freedom of movement, citizenship and Brexit in the lives of Britons resident in the European Union funded by the UK in a Changing EU. She is author of *The British in Rural France: Lifestyle Migration and the Ongoing Quest for a Better Way of Life* (2011, Manchester University Press) and co-editor of *Understanding Lifestyle Migration* (2014, Palgrave Macmillan).

Daniel Conway is a Senior Lecturer in Politics and International Relations at the University of Westminster. His research on gender, sexuality and race in South Africa includes: *Masculinities, Militarisation and the End Conscription Campaign: War Resistance in Apartheid South Africa* (2012, Manchester University Press and Wits University Press) and, with Pauline Leonard, *Migration, Space and Transnational Identities: The British in South Africa* (2014, Palgrave Macmillan). His current Leverhulme Trust Fellowship is titled: 'The Global Politics Pride: LGBTQ+ Activism, Assimilation and Resistance'.

Sophie Cranston is Lecturer in Human Geography at Loughborough University, UK. Her ESRC funded doctoral research interrogated the category of expatriate, looking at how this is produced in Global Mobility Industry, and practised among British migrants in Singapore. This research includes publications in: *Environment and Planning A* (2014), *Environment and Planning D: Society and Space* (2016), *Geoforum* (2016), *Population, Space and Place* (2017) and the *Journal of the Ethnic and Migration Studies* (2018). More recently, she has been examining the return of British migrant young people to UK Universities, funded by the RGS-IBG.

Kelly Hall is Senior Lecturer in Social Policy at the University of Birmingham. Her broad research interests include migration, ageing, health/social care and the third sector. Her recent research focuses on the migration and return of vulnerable, older British people within the EU, especially Spain, and has been published in *GeroPsych* (2016), *Ageing and Society* (2016) and *Journal of Ethnic and Migration Studies* (2016). She has recently co-designed the website www.supportinspain.info with the British Consulate in Malaga.

Katie Higgins is a post-doctoral researcher at the Department of Urban Studies and Planning, University of Sheffield. Her research explores urban inequalities, migration and privilege. Her current work focuses on wealthy elites in second-tier cities, funded by the Urban Studies Foundation. Previous work exploring British migrants living across continental Europe following the EU referendum, as well as the colonial continuities and discontinuities of British migrants living in Auckland, Aotearoa New Zealand, is published in: *Transactions of the Institute of British Geographers, Population, Space and Place, Social and Cultural Geography* and *Area*.

Priya Khambhaita is a Senior Policy Researcher at the Pensions Policy Institute (PPI). Priya has an undergraduate degree in Sociology from Cardiff University. She also has a Master's and PhD in Social Science from the University of Southampton. As well as pensions and retirement policy, her research experience includes work on the well-being of older people including the areas of health, social and economic wellness, cancer and social care.

Sarah Kunz received her PhD in Geography from University College London. Her ESRC-funded research interrogates the migration category 'expatriate', exploring its tensions and ambiguities, social histories, current usages and political effects. For this project, Sarah conducted research in Nairobi and The Hague. Sarah previously researched privileged forms of migration in Cairo and female labour migration from Bolivia to Spain. Sarah has worked as a researcher at NatCen Social Research on topics including migration, national identity and diversity.

Pauline Leonard is Professor of Sociology at the University of Southampton, where she is also Director of the ESRC South Coast Doctoral Training Partnership. She has longstanding research interests in professional migration in postcolonial contexts, and the intersections of race and ethnicity, gender and class in the construction of transnational identities. Publications include: *Expatriate Identities in Postcolonial Organizations: Working Whiteness* (2010, Ashgate); with Daniel Conway, *Migration, Space and Transnational Identities: The British in South Africa* (2014, Palgrave Macmillan); and, with Angela Lehmann, eds. *Destination China: Immigration to China in the Post Reform Era* (2018, Palgrave Macmillan).

Jenny Lloyd is Research Fellow at the University of Bedfordshire within the International Centre: Researching Child Sexual Exploitation, Violence and Trafficking. Her current research focuses on adolescent safeguarding, particularly gendered violence and young peoples' experiences of harm outside of the home. Her doctoral research explored gendered experiences of migration in Singapore and has been published in *Gender, Place and Culture* and *Area*.

Katie Walsh is Senior Lecturer in Human Geography at University of Sussex and affiliated with the Sussex Centre for Migration Research. Her work focuses on home, intimacy, and British migration. She is author of *Transnational Geographies of the Heart: Intimate Subjectivities in a Globalizing City* (2018, Wiley), and co-editor of *Transnational Migration and Home in Older Age* (2016, Routledge) and *The New Expatriates? Postcolonial Approaches to Mobile Professionals* (2012, Routledge).

Rosalind Willis is an Associate Professor in Gerontology at the University of Southampton, and is an Irish migrant to Britain. She holds a PhD in Gerontology from King's College London, an MSc in Forensic Psychology from the University of Kent, and a BSc in Psychology from Bangor University. Her research interests include: how ethnicity patterns experiences with mental health and social services; family care in later life; and the relationships between migration, culture, and identity. She has worked in mental health services and in higher education, and currently teaches social gerontology, dementia studies, and research methods.

Acknowledgements

Pauline and Katie would like to thank all the contributors for their engagement with the aims of the volume, as well as those involved in the editing and production at Routledge.

1 Introduction: British migration

Privilege, diversity and vulnerability

Pauline Leonard and Katie Walsh

Introduction

Estimates suggest there are currently 5.6 million British nationals resident outside of the United Kingdom (UK), the equivalent of one in every ten Britons, with communities of 1,000 or more in almost every country of the world (Finch, Andrew and Latorre 2010).[1] Admittedly, it is difficult to know precisely how many British nationals leave, return, or continue to reside outside of the UK each year, as collecting accurate figures on British out-migration is notoriously complex. However, in the mid-2000s, British migration drew the attention of the UK's Institute of Public Policy Research (IPPR) and the two reports published subsequently (Finch, Andrew, and Latorre 2010; Sriskandarajah and Drew 2006), as the only quantitative attempts to estimate the scale and extent of British migration, remain our source of these figures. Like others before them (e.g. Findlay 1988; O'Reilly 2000), these reports note the difficulties of data collection: British nationals rarely register with a British consulate or local government office, unless advised by the Foreign and Commonwealth Office travel guidance that they are relocating somewhere high-risk. As such, many of the estimates used in the IPPR reports rely on sources that they identify as incomplete – the International Passenger Survey has a very limited sample, passport issue and renewals may take place during visits to the UK, the Department of Work and Pensions records only the number of people who declare their overseas residence in claiming their UK state pension. That reliable estimates from census data are only available for a small number of countries only adds to this sketchiness (Sriskandarajah and Drew 2006, 8–9). Nevertheless, working with this range of sources, their estimates provide convincing evidence that the numbers of British nationals resident overseas are undoubtedly significant.

Yet, while British migration is hugely significant in statistical terms, this is the first edited volume that takes British migration as its starting point. This is perhaps surprising, given the real interest in British migration demonstrated over the last 20 years in both academic and cultural arenas, with a proliferation of book chapters, journal articles, as well as television programmes and newspaper articles exploring what it is like to be British and living abroad.

Our aim here is to bring a range of cutting-edge, in-depth qualitative studies together, including side-by-side contributions from established researchers as well as recent doctoral research projects. Doing so allows us to examine some of the diversity and commonalities among British migrants in terms of the routes and rhythms of their international mobilities and residence, lifestyles, practices, experiences, and subjectivities. Ethnographic studies have already established that individual 'British communities' in particular locations are internally diverse yet share similarities across the globe, but this collection allows us to deepen this analysis across geographically, politically and socially distinctive research sites. Chapters draw upon theoretically informed empirical research, developed in different international contexts and by researchers with different disciplinary orientations and backgrounds. We ask: What is the shape and significance of contemporary British migration? Who are today's British migrants? How might we understand their everyday lives? What can we learn about Britishness from examining how it unfolds beyond Britain? And, in the context of global and national debates about the nature of citizenships, the 'Brexit' vote[2] and deliberations surrounding mobility and freedom of movement, national, racial and ethnic boundaries, what can we learn about the contemporary British in relation to historical privileges, international relations and senses of national identity?

Our focus here on British out-migration is not only unique in that this is the first collection of this kind, but also a timely corrective to the consistent political and media attention focused on migration *into* Britain. There is intensive interest and debate surrounding people coming to Britain, yet the numbers of Britons who migrate each year, and the reasons for this, remain largely invisible – and unquestioned as a 'right' – in popular discourse. Although, as stated above, there has been some public policy focus on British out-migrants and the British 'diaspora' (e.g. Sriskandarajah and Drew 2006, Finch et al. 2010), this often tends to be romanticised as progressive: migrants are represented as being highly engaged in the communities they reside in, and as such key partners in UK public diplomacy goals (Finch, Andrew and Latorre 2010). Whether this is the case in practice remains under-examined. Academic scholars have been more critical in their approach to British migration, viewing the routes, practices, and identities of contemporary British migrants as reflecting ongoing traces of Empire (e.g. Benson and O'Reilly 2018; Conway and Leonard 2014; Fechter and Walsh 2010; Higgins 2018; Knowles 2005, 2008; Knowles and Harper 2009; Leonard 2010; Walsh 2018).

Building on existing research, in this volume we identify three significant thematic and conceptual emphases that have broader relevance for understanding British migration, namely: (1) privilege; (2) diversity; and (3) vulnerability. We have loosely organised the discussion of the chapters, in what follows here, on this basis. While it is important to recognise that many chapters speak to more than one section, this structure allows us to make a three-fold argument about contemporary British migration. The first set of chapters (from authors Benson, Abel, Cranston, Kunz) take as their central focus the critical analysis of *evidence of the race and class-based privilege of British migration and the associated terminology of 'expatriate'*. The second set of chapters (Conway and Leonard, Khambhaita and Willis, and Higgins) examine *the diversity of Britishness and mobile Britons in respect to nationality, ethnicity and race*. Finally, the chapters in the third section (Lloyd, Hall, Walsh) explore the *coexistence of vulnerabilities in accounts of the individual biographies and communities of British migrants*. As such, the volume asserts the heterogeneity and complexity of British migration. As a starting point for this assertion, we present an overview of the range of mobilities involved in contemporary British migration.

Routes and rhythms of British migrant mobilities and residence

To develop conceptual understanding of contemporary British migration, in this section we explore the range of routes and rhythms evident in the flows of today's mobile Britons. Our aim here is not only to introduce contemporary British migration to readers unfamiliar with the field, but also to contribute to research in this area by highlighting some of the connections between currently separate literatures and identifying the gaps in stories told about British migration. Reviewing existing literatures, we can identify at least five categories of British migrant: (1) settler migrants; (2) highly skilled migrants; (3) lifestyle migrants, including retirement migrants; (4) 'middling' transnational migrants; and (5) 'reverse migrants'.

The first category of British migration that we propose in this typology is *British settler migration*. From the 1880s–1960s British migration was primarily in the form of *settler migration*, with British migrants 'reinforcing an existing wider British world' and 'on the whole easily assimilated into it' (Constantine 2003, 19). They were part of a mass movement of the British population to empire destinations, especially Canada, Australia, and New Zealand, but also South Africa, India, Shanghai, Hong Kong and Singapore. These colonial societies, where the language of government and trade at least was English-speaking, continued to be among the main destinations for British settlers up until independence and beyond, attracting a range of Britons with varied skill-sets: men and women professionals, tradespersons, state functionaries (in government, the police, railway, education and health sectors) and entrepreneurs (Bickers 2010). Their migration was often heavily subsidised through 'assisted passage' or 'Ten Pound Pom' schemes, which persisted well into the 1970s and 1980s in contexts such as Australia, Canada and South Africa, seeking to shore up white presence (Hammerton and Thomson 2005; Conway and Leonard 2014). Motivations for settler migration were mixed, but were often driven by perceptions of relative poverty and/ or lack of economic opportunity at home (Leonard 2013). While forging a new material existence was not necessarily immediately any easier in the new world than the old, many persisted to craft new lives for themselves, frequently framed within a determination that they were leaving behind the repressive social structures of 'the mother country' (Elder 2007). Today these communities of settler migrants remain attractive to Britons seeking to emigrate more permanently, for instance Australia is the top destination for British migrants (Finch, Andrew and Latorre 2010). While more recent British migrants are framed as lifestyle or skilled migrants (see below), there remain in these communities some older Britons who moved themselves as settler migrants, others who were raised as the second-generation in British settler families, as well as many more whose migration is informed by the 'expatriate' practices and subjectivities of British settler migration. Sarah Kunz's chapter focuses on British settler migrants in Kenya now in later life, while Daniel Conway and Pauline Leonard's chapter draws on the stories of two Britons who attempt to position themselves against the dominant lifestyles and attitudes of British settler migrants in South Africa.

Secondly, *British professional migrants* are variously described in the wider literatures as highly skilled migrants, 'skilled transients', global talent, corporate migrants, expatriates, or elite migrants (e.g. Beaverstock et al. 2002, 2005, 2011; Beaverstock and Bordwell 2000; Findlay 1988; Findlay and Garrick 1990; Findlay, et al. 1996; Harvey 2008). Movements of the highly skilled are associated with globalisation as professionals employed in the finance, engineering, science, health, IT, law, and petroleum sectors move within transnational corporations as inter-company transferees. These British migrants have attracted the attention of human resource management studies, as well as social scientists (see Findlay and Cranston 2015 for an overview). Allan Findlay's (1988) early quantitative study of emigration identified that skilled transient migration of professional British citizens *circulating* at an international level was increasingly important from the 1960s onwards, becoming more dominant than the traditional 'settler' migration in the postcolonial period. These British migrants are most often found in global cities and key finance, IT or engineering hubs, but would also include those working for British consular services, the United Nations and other high-profile NGOs. Global flows of *temporary* highly skilled migrants were, and continue to be, dominated by men, although many relocate with their wives and children accompanying them (Coles and Fechter 2008; Findlay 1988; Hardill 1998). They may return to Britain between postings or relocate from one country direct to another, usually every 2–5 years, sometimes feeling little agency in these decisions about their movement made by 'the company'. In this volume, it is the Britons in Singapore who feature in chapters from Sophie Cranston (Chapter 4) and Jenny Lloyd (Chapter 9) who, collectively, most fit within this category of British migrant.

Lifestyle migrants, including retirees, are a third category among whom Britons are significant. For Benson and O'Reilly (2009, 608) 'lifestyle' is an analytical tool that can be of use in explaining the relocation of relatively affluent people within the developed world 'searching for a better way of life'. Most research has focused on British communities in Europe, often dominated by retirees (e.g. Benson 2011; Benson and Osbaldiston 2014; King et al.

2000; Oliver 2008; O'Reilly 2000). As O'Reilly's (2000) ethnographic study made clear, lifestyle migrants have a range of mobility patterns, from seasonal to permanent residence, and vary considerably in terms of economic resource (O'Reilly 2000). Leivestad's (2017) recent research with British residents of a caravan site in Spain certainly brings into question the idea that all lifestyle migrants are relatively privileged since they had been employed in low-paid work prior to their migration as, for instance, cleaners, healthcare assistants, mechanics, and builders. Nevertheless, their move to Spain shares in common with more affluent migrants an understanding of migration as 'a search' for a better way of life, since it is a lifestyle project that is necessarily comparative, revolving around their desire to access a favourable climate and cost of living, stronger sense of community and slower pace of life (Benson and O'Reilly 2009, 610). For the *British retirement migrant*, an enormously significant sub-group of lifestyle migrants, this lifestyle migration project is consistently framed as a strategy of active or positive ageing, part of a refashioning of identity in (early) later life (Oliver 2008). However, those researching retire-ment migration have also focused on the social care implications for those who migrate and the vulnerabilities associated with ageing (e.g. Botterill 2016; Hall and Hardill 2016; Oliver 2008). In this volume, Benson's chapter on Britons in France and Hall's chapter on British retirees in Spain fit most obviously within this vein of literature.

The almost completely distinct debates on settler migrants, highly skilled temporary migrants, and lifestyle migrants that has emerged in literatures on migration is remarkably persistent, yet empirically it is becoming harder to justify. Moreover, British migrants make decisions based simultaneously on multiple factors including both employment/income and lifestyle, something that becomes evident from their accounts in all these literatures. Furthermore, ethnographic studies that have been conducted with British migrants in par-ticular urban sites suggest a more complex picture, with settler, highly skilled migrants and lifestyle migrants in co-residence. For example, existing pub-lications on Hong Kong (Knowles and Harper 2009; Leonard 2010), South Africa (Conway and Leonard 2014) and Katie Higgins' chapter, this volume, on the British community in Auckland, Aotearea Australia provide evidence of this multiplicity (also Higgins 2018). Abel's chapter on British migration to Perth, Australia, also unsettles the conventional distinctions between the skilled and lifestyle migration literatures. It does so because the lead British migrants in this study are women and, although it is their employment as nurses and midwives that facilitates their emigration (intended by most to be relatively permanent), this also positions them outside the highly skilled labour markets of global cities which are the focus of those concerned with globalisation.

As such, we find it useful to identify here a *fourth category of British migrants: 'middling transnationals'*, to borrow Conradson and Latham's (2005) term. These British migrants are a diverse group in respect to their occupation, resource, and subjectivities (Scott 2006; Lee and Wong 2018;

Leonard 2010; 2018; Walsh 2006). We can observe, however, that their migration is more often self-initiated, in response to international recruitment initiatives, and sometimes speculative, where they might relocate before finding employment or stay on after a contract ends to seek another. These British migrants are skilled migrants too, and may include professionals, but they may also work in either lower-paid graduate sectors, especially education (including English-language tuition) and journalism, or, alternatively, have pursued non-graduate careers in the retail, hospitality, health, and tourism sectors. They may place themselves in global cities living alongside their highly skilled compatriots, but also find opportunities in other (mostly urban) locations beyond these dominant migration hubs. They vary considerably in terms of their length of residence, frequently staying longer than the number of years they initially planned and with no predetermined plan. Chapters by Conway and Leonard on British migrants in South Africa, as well as Walsh on British migrants returning from Dubai, also provide evidence of more diverse communities including this third group of 'middling transnationals'.

'Reverse migration' is a fifth category of British migration of increasing importance. This is a term that can be applied to the out-migration of British nationals with diasporic heritage, including first-generation migrants returning to their countries of origin or subsequent generations, born in the UK, reanimating the transnational social fields of their parents and grandparents (e.g. Kea and Maier 2017; Näre 2016; Potter 2005; Ramji 2006; Reynolds 2011). These 'returning' migrants are rarely understood through the framework of British migration at all. Yet, such migrants are likely to become a larger proportion of future movements out of Britain. The resurgence of 'reverse migration' by British-born citizens, moving 'back' to the country their ancestors left, also raises new questions of transnationalism and how migration is enacted 'on the ground', when a migrant may anticipate an inherited familiarity with the customs, tradition and language of the 'homeland' but have no direct experience of living there. In this volume, the chapter by Priya Khambhaita and Rosalind Willis (Chapter 7) explores their preliminary research with British-born Indian second-generation 'reverse migrants' to India, but this is certainly an understudied dimension of British out-migration. It is important to note that where these British migrants with diasporic heritage have a third country as their destination, they are already implicitly located as highly-skilled, lifestyle, or middling transnationals within the typology above. For example, British migrants with parental connections to India and Pakistan reside across the Gulf as part of the highly skilled migration noted earlier, but their migration would not be considered 'reverse migration'. It is worth noting at this point that British migrant communities are almost always theorised and researched as though they are exclusively white communities. Later in this introduction, we develop our argument that more attention should be given to the plurality of British migrants in respect to their cultural, ethnic and religious identities.

While return is implied in the 'reverse migration' discussed above, the terminology of *British return migrants* is typically reserved for those who are returning to Britain. Returnees may have left Britain as migrants themselves, perhaps multiple times, or may also be second-generation British citizens, born and/or raised overseas, who have, perhaps, repeatedly travelled 'home' with parents or to visit family, but have never lived there long-term prior to returning. Returns may be at any time and for many reasons, but are notably concentrated at some distinctive life stages/events, especially: higher education, retirement, and in the final stages of life when faced with ill health or bereavement associated with ageing (Giner-Montford, Hall and Betty 2016). Return is not positioned here as a separate category of British migrant, since it is part of all these global mobilities albeit it in varied ways. For example, many of the members of the East Africa Women's League that Sarah Kunz interviews for her chapter have lived in Kenya for most of their adult lives and can be understood as British *settler* migrants (see p. 000), yet most have not acquired citizenship. They may not intend to return, but they live with transnational imaginations sustained by the possibility of return. Some Britons who emigrate with 'forever' in mind, end up returning (e.g. Thomson 2011; Giner-Montford, Hall and Betty 2016), while some who move initially for a short period of contracted employment, may settle longer-term in one place and return reluctantly (e.g. Walsh, Chapter 11). Experiences of return are varied, shaped in part by their social status and resource (for example, compare: Knowles 2008; Rogaly and Taylor 2009; Thompson 2011). The location and meaning of home, as well as patterns of dwelling and mobility, settlement and migration, do not appear to map neatly onto different types of British migration or their geographical routes. As the chapters in this volume make evident, British migrants have a range of rhythms to their mobility and residence, the complexity of which is difficult to understand from research conducted at a particular point in their lives.

This typology is offered as a starting point only. Individual chapters in the volume illuminate the way in which communities of British nationals in particular places, as well as individuals themselves, rarely fit neatly into these categories. In time, it may become necessary to further refine the typology we propose here as increasing research illuminates further flows and communities of Britons overseas, and new trends in British migration emerge as important. In the meantime, we note that there are a number of other British global mobilities not captured by the term migration. These include, but are not limited to: the Armed Forces serving overseas; International Student Mobility (ISM) (e.g. Findlay et al. 2012); Visiting Friends and Relatives (VFR); those Britons travelling and dwelling as seasonal or year-long working holiday-makers (Clarke 2005), volunteers, and backpackers; and, finally, the population of Britons working on cruise ships and yachts who may be continually on the move outside Britain for long periods. These global mobilities exist in relation to migration as connected processes, both spatially (especially in terms of their concentration in global hubs) and temporally (especially in terms of the life-course). They are beyond the central remit of this volume, but it is useful to recognise these other British mobilities as operating alongside, and sharing partial meaning with, British migration. In respect to the meaning of British migration, one of the tensions in this typology of British migration that demands our attention is the significance of privilege in how they are framed, and how it might be related to imperial, class, and occupational status. We explore this further in the next section.

Privilege and the British 'expatriate' migrant

It is widely established that British migrants are among the most privileged of international migrants. Indeed, this is perhaps part of the reason why there has been such reticence among scholars to recognise the significance of Britons, numerically and qualitatively, within mainstream accounts of global migration. However, many in this field have argued that migration itself can be seen as a strategy for the reproduction of Briton's middle class social, cultural and economic capital, not least through the mobilisation of nationality and race as resources (Benson 2011; Benson and O'Reilly 2018; Scott 2006; Knowles 2003; Conway and Leonard 2014). While the British have clearly lost some of their political power globally, many British migrants continue to enjoy an enhanced lifestyle and status upon their migration, just as (white) colonialists did in the past. Therefore, it is important to examine the relationship between Britishness and privilege directly, something which has previously been under-theorised in studies of British migration.

As such, we start the substantive chapters of the volume with Michaela Benson's contribution (Chapter 2) in which she revisits research findings from an earlier project on British lifestyle migrants in France (Benson 2011), to propose a conceptual innovation to how we conceive of privilege among Britons living in Europe (pre-Brexit). In her study of the British residents in the Lot department of France, Benson's focus is on lifestyle migrants (see p. 000). Benson argues that it is constellations of privilege that facilitate their migration, shaping and influencing their patterns of settlement. Crucially, she suggests that privilege should be considered as an inherently and inextricably classed and racialised formation, wherein whiteness intersects with the resources of socio-economic privilege to construct *a sense of entitlement* to mobility and settlement which transcends national boundaries. However, Benson's emphasis on the *relativity* of British migrants' affluence and privilege is crucial, since it is important not to make the mistake of early whiteness studies which were somewhat guilty of essentialising and universalising whiteness, misrecognising the instabilities of whiteness as it intersects with class, ethnicity and gender to produce wide discrepancies in meanings and experience.

Gillian Abel's chapter (Chapter 3) on skilled healthcare workers emigrating to Australia, while ostensibly about a rather differently motivated group, can also be understood through notions of lifestyle and social mobility connected with the relatively privileged nature of British migration. As one quote in her chapter evokes – 'It is ours: we've got a pool! You couldn't do that in England!' For the British women she interviewed, migration brought about a change in lifestyle. Higher house prices in the UK, along with more disposable income from their new Australian employment, afforded them access to property ownership that was central to their lifestyle project. Home ownership, rather than renting, is key, Abel argues, since accessing the 'Australian Dream' for these British families, hinges on this performance of middle-class status, but also offers a marker of lifestyle distinct from that back home and a more permanent sense of settlement. As such, privilege is constituted not only in contrast to how other groups might be positioned within Australia, but also in relation to a transnational social field.

The word 'expatriate' has often been associated with privilege in both popular usage and academic discussion. It acts as a common nomenclature for (white) 'Western' skilled professionals who live abroad for a temporary period of time (Fechter 2007; Knowles and Harper 2009; Kunz 2017; Leonard 2010). Those researching British migration have been among those critically examining the production and function of this terminology, so the volume continues with two chapters that make further contributions to these debates. Sophie Cranston builds on her previous work in which she has demonstrated the key role played by the global mobility industry in producing imaginations and understandings of difference through which Britons *learn* to be an expatriate in relation to a particular location (Cranston 2016a, 2016b). Rather than 'expatriate' being a conceptual term to differentiate between different kinds of migrant, Cranston shows how it operates within destination services and intercultural management that prepare the migrant for life in Singapore as 'a discursively produced object to which individuals can orient themselves' (Cranston 2016a, 3). In her chapter in this volume, Cranston takes this argument further by shifting the focus away from the *place* of migration, to the *relationships* produced within and between the British migrants who live there. The respondents in her study position themselves very differently in terms of the identity of 'expatriate'. Since it is a term loaded with cultural baggage, British migrants either gravitate towards it or energetically resist. For those who rejected the label of expatriate, the term was associated with a performance of Britishness which harked back to the worst excesses of colonialism: segregated and privileged Britons who held little interest in engaging with Singaporean culture and politics. In contrast, Cranston describes how an alternative performance of Britishness is demonstrated by the '*Ang-Mohporians*', who attempt to distance themselves from the representations, and practices, associated with the traditional expatriate. and hold a more local type of belonging. Her account is revealing in that it underscores how senses of difference and distinction infuse British national identities: who we *are* being a carefully crafted negation against who we *are not*.

Sarah Kunz also explores the category of 'expatriate' as it is used and negotiated by British migrants themselves, revealing its complexity, instability, and how it is met with both enthusiasm and aversion. Her study involves interviews with members of the East Africa Women's League, an organisation for European women, in Nairobi. Crucially, she reveals the work that the label 'expatriate' can do in terms of whiteness, arguing that 'expatriate' is mobilised uncritically and habitually against an 'African Other' informed by colonial imaginations and histories. Kunz demonstrates that 'expatriate' is also used by some to reference their status as non-citizens: for some, this was narrated as a strategic choice drawing on the privilege of transnational resource and potential mobility or return, whereas for others it reflected anxieties concerning a changing status in relation to the postcolonial state. Kunz argues that the ability to hold onto attachments to Britishness rests in on-going coloniality and the continuing economic and political power of the global North.

The volume brings attention then, especially in these four chapters, to the relative privilege of British migrants in varied geographical contexts and the way in which, for some British communities, the terminology of 'expatriate' is part of the everyday making and marking of this privilege. It is important to remind ourselves that British migrants may reside and socialise largely in British communities or broader migrant communities, which may in turn be based on Europeanness, whiteness (including US, Australian, New Zealanders and white South Africans for example), or communities of English-speaking highly skilled migrants (that may also encompass professionals for whom English is not their first language).[3] Indeed, locating themselves within these communities is part of the strategy of accessing privilege through social networks and association with a certain set of lifestyles. As such, a broader critical literature on 'expatriate' migration is also useful for understanding the shape and affiliations of their lives beyond the national (e.g. Farrer 2010; Fechter 2007; Fechter and Walsh 2010), including that on aid workers (Fechter 2013). The empirics in the chapters by Cranston and Kunz start to unpack some of the internal heterogeneity of British communities in particular places, by showing how Britons distinguish between themselves and other Britons, as well as how they vary in terms of identification with not only the label itself but also (what are perceived to be) expatriate practices and lifestyles (themselves arising in place as geographically distinct).

It is also important to remind ourselves that not *all* British migrants are set on accessing privilege through migration. Migration involves ongoing processes of adjustment and modification according to context and shifting conditions (Amit and Knowles 2017) and diversity is revealed within the ways in which people position themselves in relation to privilege. The volume now turns further to explore the heterogeneity of British migrant subjectivities, through the analysis of the diversity of British migrants in terms of nationality, ethnicity, and political and social orientation, as well as the vulnerabilities brought about by the intersection of gender, age and class with Britishness.

'Britishness' and the diversity of mobile Britons

It is perhaps a very British characteristic to shy away from looking too closely at Britishness and 'The British'. National anxieties about regionalism, historical imperialism, international relations, immigration and the failure of multiculturalism can coalesce to deflect internal examination and encourage short-hand, 'unified' imaginations of British national identity and citizenship (Alibhai-Brown 2000; Kumar 2000; Gilroy 2002; McCrone and Bechhofer 2015; Bhambra 2017). Scholars too raise legitimate concerns that, in the very act of thinking and talking about Britishness, it will become reified as a fixed category of experience, 'as a monolith, in the singular, as an "essential something"' (Fine et al. 1997, xi). Indeed, in many of its more recent iterations in the media and popular debate, Britishness, or more frequently, Englishness, is often elided with 'conservative mind-sets and the support of militant nationalism, war, and outmoded notions of class and race' (Knuth 2012, vii). Nevertheless, not least in the wake of 'Brexit', academics, policymakers and cultural commentators have continued to debate the nature of Britishness and British identity, its relationships to the devolved nations of Scotland, Wales and Northern Ireland as well as Europe, multi-nationalism, ethnic pluralisation and immigration (Parekh 2006; Tilley and Heath 2007; Modood 2016; Bhambra 2017).

The vibrant and ever-growing field of research into British migration has added significantly to our understanding of the complexities and diversity of British national identity. The intersections of class, gender and race, in interplay with post/colonial dis/continuities, are widely revealed as integral to the multiplicity of British identities and the different ways these are negotiated across time and space. By drawing together a range of the most contemporary research conducted with British migrants from different backgrounds and in different national contexts, this collection makes an important contribution to the examination of the meanings of contemporary Britishness as it plays out beyond the geographical shores of Britain, as well as to what these may tell us about Britishness at 'home'. What is resoundingly clear is that the social structural factors which act to diversify domestic Britons can multiply even further in migratory contexts, as they intersect with the specific political histories and social landscapes of other locales. All the while, the legacies of empire and imperial modes of governance remain as powerful resources for identity construction and, as often as not, hierarchical organisation, albeit interwoven with new meanings and performances drawn from both the local and migrant communities. Individual migrants are therefore shaped by the possibility of multiple and shifting subject positions, and the consequent formation and reformation of identities is a continuous process, accomplished through actions and words rather than any fundamental essence of national character. The next three chapters in the collection explore these diverse identity-making processes in fine detail.

Daniel Conway and Pauline Leonard's discussion in Chapter 6 offers a counterpoint to the research on British out-migration that predominantly focuses on the lives, experiences and identity-making processes of white, privileged migrants. While this has undoubtedly led to rich, productive and critical analyses of the constructions of British nationality, in intersection with whiteness, in migratory contexts, there remains a gap in our knowledge of British migrants of diverse and mixed racial and ethnic backgrounds. Drawing in-depth on the stories of Andrew and Caroline, two white British migrants who 'married out': that is, each married a black and a 'Coloured' (mixed race) South African and built new lives with their partners' extended families, the chapter demonstrates how interracial marriage can force migrants to address what are often, for British migrants, new questions of identity, social and political affiliation and lifestyle. The chapter introduces the concept of 'transnational normativity' to conceptualise the ways in which, historically, white British migration to South Africa has been primarily and predominantly fuelled by desires for, and promises of, upward social and economic mobility (Conway and Leonard 2014). In the post-apartheid era, many continue to enjoy life in South Africa because of these material and social benefits. Through and within this, a race- and class-based British 'way of life' can be easily established in the new national context, with transnational performances and relations maintained. However, there are British migrants who transgress this normativity, deliberately attempting to turn away from the apparatus, and the associated privileges and identities, normally pertaining to the British migrant community in South Africa. While the analysis reveals that some of the benefits which accrue to whiteness and Britishness continue to inflect the 'transgressive' British migrants' racial beliefs and practices, albeit less overtly, the discussion of Andrew and Caroline's lives reveals the emerging pattern of identity-making choices and the importance of what Caroline Humphrey calls 'decision-events' to 'denote a moment out of the ordinary in which people "open themselves to a radically different composition of the self, a switch that has a lasting effect" (2008: 371, quoted in Amit and Knowles 2017, 170). The narratives demonstrate how identities, political orientations and experiences produced at home may be drawn upon to position themselves as 'different' in the South African context. In the process, their stories challenge the dominant structures of British migrant identity-making, adding further layers of nuance to what it may mean to be British.

Priya Khambhaita and Rosalind Willis' chapter (Chapter 7) also explores the negotiation of differing cultures in the construction of new identities and lifestyles. Their chapter draws on qualitative research conducted with second and third generation British-born Indians (BBIs) who migrate to India, which is now perceived to offer a number of lifestyle benefits to the Indian diaspora, such as better investment opportunities, improved living standards in terms of accommodation, a wider choice of schools for children and an improved context for caring for ageing parents and grandparents. Most BBIs recognise that their migration to India impacts on their ageing relatives in the UK and, as such, would plan for them to join them. Not only does an Indian (middle-class) lifestyle offer new privileges of private healthcare, domestic help and a more relaxed lifestyle, the collectivist culture of Indian society is seen to more easily facilitate extended family living, supporting the notion of the domestic home as a source of reciprocity between generations. However, whereas within the domains of home and family BBIs may identify more with their Indian heritage than their British, in other areas returnees were found to identify more with being British. In terms of interactions with locals such as colleagues, and differences in personal conduct and manners, respondents faced challenges in terms of accepted modes of communication and behaviour which reinforced their identification as British.

Katie Higgins further examines the varied experiences and expressions of Britishness in her discussion in Chapter 8. She rightly notes that there has been relatively little attention to the variation between Britain's constituent nations when researching the British as a migrant group, which not only risks essentialising 'the British' as a homogeneous national group but can, as often as not, mean an over-emphasis on the English as the majority national culture. Her focus is on first-generation migrants in the context of Aotearoa New Zealand, and her exploration of the dimensions of Britishness for English, Welsh and Scottish migrants produces strong evidence that these national belongings also matter. She notes how English participants tended to distance themselves from overt displays of patriotism and compatriots who were overly focused on recreating Britain socially and culturally, particularly sensitive to a lingering stereotype of 'whinging poms' and cultural insensitivity. In contrast, those who came from nations which might draw on a 'Celtic' identification were more likely to proclaim a national (Welsh/Scottish) identity with some celebratory pride: in part to distinguish themselves from the English who are often (and incorrectly) positioned more closely with the making of empire. Yet, contradictorily, for some of these participants 'Britishness' offered a more ambivalent and flexible identity, which could be drawn upon as a resource to indicate a cosmopolitan-multiculturalism: as Charles explains, Britishness 'is a nice expression because it, sort of, it almost captures the essence of multiculturalism'.

These three chapters bring attention to some interesting themes that emerge from the research in spite of the wide range of geographical, social and political contexts. First is the reinforcement of the diversity, heterogeneity and somewhat ephemeral nature of Britishness as a national identity. Together, the chapters reveal its constantly changing iterations, as it is made and remade by different groups of migrants. Second, is the individual level of heterogeneity: Britishness is not only a fractured and sometimes contested position between individuals but within individuals, who may, in one moment and in one context, display one aspect of identity and, in another, another (perhaps competing) construct of identity. Third, is that the political and social debates around the nature of British identity engaging Britons at home clearly continue to resonate in identity-making processes in migratory contexts. Far from, for example, issues of multiculturalism being 'in retreat' (Meer and Modood 2009) as Britons journey to new lands, the deliberations and careful negotiations of these British migrants from diverse backgrounds would appear to reflect an ongoing engagement with questions of cultural diversity and how to position oneself within these. In the ways that migrants' subjectivities and performances are confronted anew by local and indigenous identities and social relations in the new migratory context, what Britishness means in terms of relations to post/colonialism and empire, integration, social cohesion and relations to 'difference' continues to be worked out on a global stage through the micro-level, day-to-day decisions, actions and relations of individual migrants themselves.

Vulnerability and British migration

One of the problems researchers of British migration face is how to humanise the individuals taking part in their research in a context where privileged migrants are, at best, stereotyped and, at worst, vilified. Katie remembers presenting her PhD research about British migrants in Dubai to academic audiences who asked how she survived hanging out with such 'airheads' when she was talking about people who had become her friends. Pauline has also experienced critique as to why she was even researching white South Africans who 'are all racists'. British migration also has an ambiguous place in the imaginaries of migration that circulate in the British media. For example, O'Reilly (2000) notes that interest has been evident from the 1990s in the number of television documentaries, soap operas, and property programmes focused on the British in Spain, but these media outputs need to be unpacked due to their stereotyping of the British as:

> either upper-class colonial style, or lower class, mass tourist style expatriates searching for paradise, living an extended holiday in ghetto-like complexes, participating minimally in local life or culture, refusing to learn the language of their hosts, and generally creating an England in the sun.
>
> (O'Reilly 2000, 6)

Desiring, as we do, to bring a critical perspective on British migration, makes the task of analysis and representation even more difficult. Where most migrants occupy, rightly, an empathetic space in academic accounts, researchers of British migration often note that their participants are denied this as a consequence of their relative privilege. Yet, as our understanding of British lives abroad deepens, it becomes evident that some Britons feel a lack of agency as they navigate global migrant labour regimes and citizenship from varied biographical and gendered positions, carrying different financial resources, and embodying or holding skills that cannot always negate the precariousness of employment in contemporary neoliberal economies. Consequently, while it is absolutely vital to theorise and highlight the privilege discussed in the previous section, it is not sufficient. To understand Britishness abroad, we must also consider the vulnerabilities of some Britons at some times.

In this vein, existing work on British migration has been influenced by theoretical perspectives that examine Britishness as racialised and classed in co-constitution with other social identifications which may work to undermine or unsettle the privilege of a white British migrant subjectivity. Much of the work on highly skilled migrants in global cities, for example, has applied an analytical framework of gender to complicate our understanding of whiteness in postcolonial urban spaces (e.g. Leonard 2008; Knowles and Harper 2009; Walsh 2007, 2011). The experience and subjectivities of British women are, of course, very diverse, but there is clear evidence that they are strongly shaped by marital status. Research has demonstrated that single British women frequently inhabit distinct social spaces from married Britons, socialising with colleagues and friends rather than in family or couple units (Walsh 2018; Willis and Yeoh 2007). Married British women face quite different opportunities and challenges since their own movement is most often wrapped up with a family or household move with their husband being the lead migrant. While couples may both be equally educated and have professional careers in the UK, the prioritisation of the male partner's career through international mobility remains a social norm for households attempting to navigate persistent gender inequalities that work at multiple levels (Hardill 1998; Leonard 2008; Walsh 2007). Existing scholarship has consistently demonstrated that the privilege of British married women is far from straightforward in terms of their felt experience, given their position within British communities as non-working wives.

The chapter in this volume by Jenny Lloyd (Chapter 9) takes these debates further through her innovative theoretical analysis of how British women feel about their embodiment in Singapore. Informed by feminist theories of embodiment and critical studies of body size, Lloyd argues that the body is a significant part of the way in which British women living in Singapore make sense of their everyday gendered experience of postcolonial encounters in the city. Rather than *feeling* privileged, the women Lloyd interviews instead reveal their anxieties about body size and sexuality. Indeed, the meaning of their classed and racialised migrant status is revealed to be complex and contingent as Lloyd presents startling confessions of everyday feelings of insecurity resulting from the need to navigate a postcolonial 'contact zone' away from home in which they read their own bodies as 'frumpy and fat'. As the women navigate clothes shopping and the relational construction of their body size, they also negotiate the geographically contingent construction of femininity and not fitting in. Lloyd traces how their emotional resistance to experiencing their own bodies as problematic often leads them to participate in the stereotyping of Singaporean women as part of a wider construction of racialised and gendered stereotypes about 'Asian' women. Indeed, Lloyd argues that in the contemporary Singaporean migration context, difference is not discussed through overt racial narratives but instead through conversations about clothing sizes and styles. Crucially, Lloyd's work furthers our understanding of the gendered experience of highly skilled migration or 'mobile professionals' (see Coles and Fechter 2008 for further discussion) and brings further insight into the operation of entitlement and privilege by its attention to moments of vulnerability. While Lloyd's approach to embodiment and gender is specifically about body size, it makes a broader contribution to the literatures on British migration by reminding us of the necessity of understanding migration as embodied. In doing so, Lloyd connects with a broader agenda in migration studies to more fully understand migrant subjectivities and invites us to explore the possibilities for studying British migration through the lens of embodiment.

While gender has perhaps been the key social identification to gain attention in studies of British migrants in global cities, the literature on lifestyle migrant communities in Southern Europe has identified ageing as another aspect of embodiment that intersects with nationality due to the large number of retirees and older Britons (e.g. Oliver 2008). In this volume, Kelly Hall's chapter (Chapter 10) examining resilience is revealing of the impact of ageing in unsettling the assumed privilege of white middle-class migrants. She argues that when Britons reach the 'fourth age', a decline in health often generates need for additional care and support, but many of the British households in her sample did not have family members in Spain and their children lived in the UK. While they were often very resilient individually, as couples, and through their friendship and social networks within the British community, in times of crisis these Britons regretted that support from family was limited by their geographical separation. We also know from existing studies that many Britons decide in later life to return to the UK due to ill-health or bereavement, sometimes in spite of their previous intentions to stay (Giner-Montford, Hall and Betty 2016; Oliver 2008).

The final chapter in the volume continues with this focus on ageing, but turns instead to the vulnerabilities associated with retirement and return. Katie Walsh selects three narratives from a project with British return migrants, to explore the experiences of returning from the UAE. She uses these accounts to demonstrate how British migrants, in spite of their privilege as skilled migrant workers in Dubai, experience vulnerability when they enter their sixties and approach formal retirement age. Irrespective of their considerable length of residence (the interviewees had all lived in Dubai over decades), the right to residence for migrants of all nationalities and occupations is connected to their working visas, a result of the *Kafala* sponsorship system through which migration is managed in the Gulf states. This creates a sense of return being inevitable and a temporariness to settlement in Dubai, even for those who have great attachment to the city. Walsh demonstrates that the impact of chronological age is not absolute, since Britons can use various capitals (social, wealth, occupational) to extend their stay and navigate their return, but that these are differentially available. This brings diversity to the experience of retirement, even among middle class Britons returning from one city.

Together the chapters in this section of the volume (by Lloyd (Chapter 9), Hall (Chapter 10) and Walsh (Chapter 11)) further remind us of the importance of unpacking Britishness to explore how intersecting social identifications, such as gender and age, might shape the production of British migrant identities. It is necessary to extend these efforts by exploring more fully the impacts of class, sexuality, age, gender, and generation on the production of British migrants' subjectivities. Indeed, this is one of the ways we must seek to humanise the subjects of any research on British migrants. Recognising the full complexities of their biographical resources and social positions as British migrants navigate the varied routes, residence, and mobilities of their projects of international migration, is a necessary under-taking if we are to adequately examine and understand contemporary British migration.

Conclusion: an agenda for researching British migration

As editors, we are keenly aware that this volume does not present a comprehensive over-view of contemporary British migration. This was not our aim. More particularly, at the beginning of this Introduction, we asked what a focus on British migration might enable us to learn about the contemporary British in relation to historical privileges, international relations and senses of national identity. Through its curation of some of the most contemporary research on British migration, and as the topography of British migration is etched a little more finely, the volume enables us to make some headway in addressing these questions. We learn more about the ongoing resonance of the historical specificities of space in its constituent practices in the production of Britishness and British identity-making, as well as the integral relationship to other national identities in its meanings and performances. We learn that while British identity almost ubiquitously delivers privilege, this is not consistently experienced or wanted. The heterogeneous patterns of, or orientations towards, Britishness are delved into with rigorous scrutiny and through original lenses such as the body, political orientation and ageing. In the process, we are reminded that the British may not only be privileged but also vulnerable, as spatial distance can diminish important resources such as family care. While the collection has added further complexity to the debates about British identity and what it means to be British, demonstrating both its reliance on everyday routines, regularities and conventions as well as its fluidity, recalibrations and ongoing dynamism, it has also provided the opportunity for us to pause and reflect on what would be productive lines of enquiry to further develop understanding of British migration. The notable absence of research about non-white British migrants is a key aspect to address. There is also a real knowledge gap in relation to the experiences of children and young people, including: firstly, those travelling with their parents in successive postings; secondly, those returning to the UK for their schooling or higher education; and, thirdly, those who are born or settle longer-term abroad as part of a second-generation of relatively permanent British emigrants. There is also a curious absence of discussion of the role of religion in regard to the formation and operation of British social networks, communities and identities overseas, something which appears in stark contrast to the focus of much research on migrants resident in the UK. In some contrast to much of this research, it is also evident that most research on British migrants has concentrated less on exploring the transnationalism of their everyday lives, such that the transnational practices, transnational social fields, and transnational families that constitute British communities overseas remain under-examined.

Notes

1 This does not include all those eligible for a British passport, nor everyone who may claim British ancestry or ethnicity.
2 In 2016, the UK voted to leave the European Union, coined as 'Brexit'. It is scheduled to depart on Friday 29 March 2019 and, at the time of publication, the detail of terms of leaving are still under negotiation.
3 The research by Sarah Kunz and Jenny Lloyd in this volume included other nationalities in their sample and authors were asked to analyse the findings in relation to their British participants (not simply remove those who weren't British) for their contributory chapters.

References

Alibhai-Brown, Y. 2000. *Who Do We Think We Are? Imagining the New Britain.* London: Allen Lane.
Amit, V. and Knowles, C. 2017. 'Tacking: improvising and navigating mobilities and everyday life'. *Theory Culture and Society*, 34(7–8): 165–179.
Beaverstock, J. 2005. 'Transnational elites in the city: British highly-skilled inter-company transferees in New York City's financial district'. *Journal of Ethnic and Migration Studies*, 31(2): 245–268. doi:10.1080/1369183042000339918.
Beaverstock, J. 2011. 'Servicing British expatriate "talent" in Singapore: exploring ordinary transnationalism and the role of the "expatriate" club'. *Journal of Ethnic and Migration Studies*, 37(5): 709–728. doi:10.1080/1369183X.2011.559714.
Beaverstock, J. and Boardwell, J. 2000. 'Negotiating globalization, transnational corporations and global city financial centres in transient migration studies'. *Applied Geography*, 20(3): 277–304.
Beaverstock, J., Taylor, P. and Smith, R. 2002. 'Firms and their global service networks'. In *Global Networks: Linked Cities*, edited by Saskia Sassen, 93–116. London: Routledge.
Benson, M. 2011. *The British in Rural France: Lifestyle Migration and the Ongoing Quest for a Better Way of Life.* Manchester: Manchester University Press.
Benson, M. and O'Reilly, K. 2009. 'Migration and the search for a better way of life: a critical exploration of lifestyle migration'. *The Sociological Review*, 57(4): 608–625.

Benson, M. and O'Reilly, K. 2018. *Lifestyle Migration and Colonial Traces in Malaysia and Panama*. London: Palgrave Macmillan.

Benson, M. and Osbaldiston, N. eds. 2014. *Understanding Lifestyle Migration: Theoretical Approaches to Migration and the Quest for a Better Way of Life*. Basingstoke: Palgrave.

Bhambra, G. 2017. 'Locating Brexit in the pragmatics of race, citizenship and empire'. In *Brexit: Sociological Responses*, edited by W. Outhwaite. London: Anthem Press.

Bickers, R. ed. 2010. *Settlers and Expatriates: Britons over the Seas*. Oxford, UK: Oxford University Press.

Botterill, K. 2016. 'Discordant lifestyle mobilities in East Asia: privilege and precarity of British retirement in Thailand'. *Population, Space and Place*. https://doi.org/10.1002/psp.2011.

Clarke, N. 2005. 'Detailing transnational lives of the middle: British working holiday makers in Australia'. *Journal of Ethnic and Migration Studies*, 31(2): 307–322.

Coles, A. and Fechter, A.-M. eds. 2008. *Gender and Family among Transnational Professionals*. New York: Routledge.

Conradson, D. and Latham, A. 2005. 'Transnational urbanism: attending to everyday practices and mobilities'. *Journal of Ethnic and Migration Studies*, 31: 227–233. doi:10.1080/1369183042000339891.

Constantine, S. 2003. 'British emigration to the Empire-Commonwealth since 1880'. In *The British World. Diaspora, Culture and Identity*, edited by C. Bridge and K. Fedorowich, 16–35. London: Frank Cass.

Conway, D. and Leonard, P. 2014. *Migration, Space and Transnational Identities: The British in South Africa*. Basingstoke: Palgrave Macmillan.

Cranston, S. 2016a. 'Producing migrant encounter: learning to be a British expatriate in Singapore through the Global Mobility Industry'. *Environment and Planning D: Society and Space*, 34(4): 655–671.

Cranston, S. 2016b. 'Imagining global work: producing understandings of difference in easy Asia'. *Geoforum*, 70: 60–68.

Elder, C. 2007. *Being Australian: Narratives of National Identity*. Crows Nest, Australia: Allen and Unwin.

Farrer, J. 2010. '"New Shanghailanders" or "new Shanghainese": Western expatriates' narratives of emplacement in Shanghai'. *Journal of Ethnic and Migration Studies*, 36: 1211–1228.

Fechter, A., 2007. *Transnational Lives: Expatriates in Indonesia*. Aldershot: Ashgate.

Fechter, A. 2013. *The Personal and the Professional in Aid Work*. London: Routledge.

Fechter, A. and Walsh, K. 2010. 'Introduction to Special Issue: "examining 'expatriate' continuities: postcolonial approaches to mobile professionals"'. *Journal of Ethnic and Migration Studies*, 36(8): 1197–1210. ISSN 1369–1183X.

Finch, T., Andrew, H. and Latorre, M. 2010. *Global Brit: Making the Most of the British Diaspora*. UK: Institute of Public Policy Research.

Findlay, A. 1988. 'From settlers to skilled transients: the changing structure of British international migration'. *Geoforum*, 19(4): 401–410.

Findlay, A. and Cranston, S. 2015. 'What's in a research agenda? An evaluation of research developments in the arena of skilled international migration'. *IDPR*, 37(1): 17–31. doi:10.3828/idpr.2015.3.

Findlay, A. and Garrick, L. 1990. 'Scottish emigration in the 1980s: a migration channels approach to the study of skilled migration'. *Transactions of the Institute of British Geographers*, 15(2): 177–192.

Findlay, A., King, R., Smith, F., Geddes, A. and Skeldon, R. 2012. 'World class? An investigation of globalisation, difference and international student mobility'. *Transactions of the Institute of British Geographers*, 37(1): 118–131. doi.org/10.1111/j.1475-5661.2011.00454.x.

Findlay, A., Li, F., Jowett, A. and Skeldon, R. 1996. 'Skilled international migration and the global city: a study of expatriates in Hong Kong'. *Transactions of the Institute of British Geographers*, 21(1): 49–61.

Fine, M., Weis, L., Powell, C. and Mun Wong, L. eds. 1997. *Off-White: Readings in Race, Power and Society.* New York: Routledge.

Gilroy, P. 2002. *There Ain't No Black in the Union Jack: The Cultural Politics of Race and Nation.* London: Routledge [first published in 1987].

Giner-Montford, J., Hall, K. and Betty, C. 2016. 'Back to Brit: retired British migrants returning from Spain'. *Journal of Ethnic and Migration Studies*, 42(5): 797–815.

Hammerton, A.J. and Thomson, A. 2005. *Ten Pound Poms: Australia's Invisible Migrants.* Manchester: Manchester University Press.

Hardill, I. 1998. 'Gender perspectives on British expatriate work'. *Geoforum*, 29(3): 257–268.

Hall, K. and Hardill, I. 2016. '*Retirement migration, the "other" story: caring for frail elderly British citizens in Spain'.* Ageing and Society, 36(3): 562–585.

Harvey, W. 2008. 'The social networks of British and Indian expatriate scientists in Boston'. *Geoforum*, 39: 1756–1765. doi:10.1016/j.geoforum.2008.06.006.

Higgins, K.W. 2018. 'Lifestyle migration and settler colonialism: the imaginative geographies of British migrants to Aotearoa New Zealand'. *Population, Space and Place*, 24(3).

Kea, P. and Maier, K. 2017. 'Challenging global geographies of power: sending children back to Nigeria from the U.K. for education'. *Comparative Studies in Society and History*, 59(4): 818–845. ISSN 0010–4175.

King, R., Warnes, T. and Williams, A. 2000. *Sunset Lives: British Retirement Migration to the Mediterranean.* Oxford: Berg.

Knowles, C. 2003. *Race and Social Analysis.* London: Sage.

Knowles, C. 2005. 'Making whiteness: British lifestyle migrants in Hong Kong'. In *Making Race Matter: Bodies, Space and Identity*, edited by C. Knowles and C. Alexander. Basingstoke: Palgrave Macmillan.

Knowles, C. 2008. 'The landscape of post-imperial whiteness in rural Britain'. *Ethnic and Racial Studies*, 31(1): 167–184. ISSN 0141–9870.

Knowles, C. and Harper, D. 2009. *Hong Kong: Migrant Lives, Landscapes, Journeys.* Chicago: University of Chicago Press.

Knuth, R. 2012. *Children's Literature and British Identity: Imagining a People and a Nation.* Plymouth: Scarecrow Press.

Kumar, K. 2000. 'Nation and empire: English and British national identity in comparative perspective'. *Theory and Society*, 29: 575–608.

Kunz, S. 2016. 'Privileged mobilities: locating the expatriate in migration scholarship'. *Geography Compass*, 10(3): 89–101.

Lee, M. and Wong, T. 2018. 'From expatriates to new cosmopolitans? Female transnational professionals in Hong Kong'. In *Destination China: Immigration to China in the Post-Reform Era*, edited by A. Lehmann and P. Leonard. Singapore: Palgrave Macmillan.

Leivestad, H. 2017. 'Campsite migrants: British caravanners and homemaking in Benidorm'. *Nordic Journal of Migration Research*, 7(3): 181–188. https://doi.org/10.1515/njmr-2017-0022.

Leonard, P. 2008. 'Migrating identities: gender, whiteness and Britishness in post-colonial Hong Kong'. *Gender, Place and Culture*, 15: 45–60. doi.org/10.1080/09663690701817519.

Leonard, P. 2010. *Expatriate Identities in Postcolonial Organizations: Working Whiteness.* Farnham: Ashgate Publishing.

Leonard, P. 2013. 'Making whiteness work in South Africa: a translabour approach'. *Women's Studies International Forum*, 36(Jan–Feb): 75–83.

Leonard, P. 2018. '"Devils" or "superstars"? Making English language teachers in China'. In *Destination China: immigration to China in the Post-Reform Era*, A. Lehmann and P. Leonard. Singapore: Palgrave Macmillan.

McCrone, D. and Bechhofer, F. 2015. *Understanding National Identity.* Cambridge: Cambridge University Press.

Meer, N. and Modood, T. 2009. 'The multicultural state we're in: Muslims, "multiculture", and the Civic re-balancing of British multiculturalism'. *Political Studies*, 57(1): 473–497.

Modood, T. 2016. 'Book review: David McCrone and Frank Bechhofer, *Understanding national identity'*. *Sociology*, 50(6): 1202–1204.

Näre, L. 2016. 'Home as family: narratives of home among ageing Gujaratis in the UK'. In *Transnational Migration and Home in Older Age*, edited by K. Walsh and L. Näre, 1st edn, 50–60. London: Routledge.

Oliver, C. 2008. *Retirement Migration: Paradoxes of Ageing.* London: Routledge.

O'Reilly, K. 2000. *The British on the Costa del Sol.* London: Routledge.

O'Reilly, K. 2018. 'The British on the Costa Del Sol twenty years on: a story of liquids and sediments'. *Nordic Journal of Migration Research*, 7(3): 139–147.

Parekh, B. 2006. *Rethinking Multiculturalism*, 2nd ed. Basingstoke: Palgrave Macmillan.

Potter, R. 2005. '"Young, gifted and back": second-generation transnational return migrants to the Caribbean'. *Progress in Development Studies*, 5(3): 213–236. https://doi.org/10.1191/1464993405ps114oa.

Ramji, H. 2006. 'British Indians "returning home": an exploration of transnational belongings'. *Sociology*, 40(4): 645–662. doi:10.1177/0038038506065152.

Reynolds, T. 2011. 'Caribbean second-generation return migration: transnational family relationships with "left-behind" kin in Britain'. *Mobilities*, 6: 535–551. https://doi.org/10.1080/17450101.2011.603946.

Rogaly, B. and Taylor, B. 2009. *Moving Histories of Class and Community.* Basingstoke, Hampshire: Palgrave Macmillan.

Scott, S. 2006. 'The social morphology of skilled migration: the case of the British middle class in Paris'. *Journal of Ethnic and Migration Studies*, 32(7): 1105–1129. http://dx.doi.org/10.1080/13691830600821802.

Scott, S. 2007. 'The community morphology of skilled migration: the changing role of voluntary and community organisations (VCOs) in the grounding of British identities in Paris (France)'. *Geoforum*, 38: 655–676. https://doi.org/10.1016/j.geoforum.2006.11.015.

Sriskandarajah, D. and Drew, C. 2006. *Brits Abroad: Mapping the Scale and Nature of British Emigration.* London: Institute for Public Policy Research.

Thomson, A. 2011. *Moving Stories: An Intimate History of Four Women Across Two Countries.* Manchester: Manchester University Press and UNSW Press.

Tilley, J. and Heath, A. 2007. 'The decline of British national pride'. *The British Journal of Sociology*, 58(4): 661–678.

Walsh, K. 2006. '"Dad says I'm tied to a shooting star!": grounding (research on) British expatriate belonging'. *Area*, 38(3): 268–278. ISSN 0004–0894.

Walsh, K. 2007. 'Travelling together? Work, intimacy and home amongst British expatriate couples in Dubai'. In *Gender and Family Among Transnational Professionals*, edited by A. Coles and A.-M. Fechter, 63–84. New York andLondon: Routledge.

Walsh, K. 2011. 'Migrant masculinities and domestic space: British home-making practices in Dubai'. *Transactions of the Institute of British Geographers*, 36(4): 516–529. ISSN 0020–2754.

Walsh, K. 2018. *Transnational Geographies of the Heart: Intimate Subjectivities in a Globalizing City*. UK: Wiley Blackwell.

Willis, K. and Yeoh. B. 2007. 'Coming to China changed my life: gender roles and relations among single highly-skilled migrants'. In *Gender and Family Among Transnational Professionals*, edited by A. Coles and A.-M. Fechter, 211–232. London: Routledge.

2 Constellations of privilege

The racialised and classed formation of Britons living in rural France

Michaela Benson

Introduction

Within the context of a volume on British migration, considerations of how privilege is constituted have particular relevance and salience. Such considerations bring into focus the shared histories upon which British emigrations are grounded, providing a conceptual framework that has value beyond understandings of those British citizens resident in Europe. In bringing in the lens on colonialism and the social production of whiteness that is common in accounts of British migration to former outposts of the British empire (see for example Leonard 2010; Coles and Walsh 2010; Knowles and Harper 2009), the reconsideration of privilege presented in this chapter allows for consideration of the (dis)continuities between these migrations further afield (see for example Fechter and Walsh 2010) and those closer to home in Europe. The June 2016 British referendum on their membership of the European Union makes this a particularly timely endeavour given that Britain's future outside Europe has undoubted consequences for those Britons who live and work in Europe, the so-called 'Brexit' an opportune moment to consider the traces of colonialism in Britain's and Britons' relationships with Europe.[1]

This chapter starts by re-visiting rural France, setting up my ethnographic research with Britons living in the Lot. It then introduces the lifestyle migration framework, highlighting the key conceptual and theoretical framing this offers and the significance of thinking relationally about affluence and privilege within a broader sociological project of 'undoing' privilege. From the discussion of British migration as a form of middle-class reproduction, it turns to the consideration of how whiteness at its intersections with Britishness is reproduced. In this way, it highlights the inherently postcolonial legacies at play in the structuring of privilege that facilitates such migrations, framing the imaginings of people and place within the destination. As I highlight, in examining the reproduction of the British middle classes on French soil it is also important to acknowledge and deconstruct this as a process that socially produces whiteness, and through which legacies of empire systemically persist.

Introducing the British residents of the Lot

From the limestone cliffs in the east of the department, through to the rolling farmland and vineyards to the west, the Lot offers picturesque and awe-inspiring scenery, a feature of the environment that became pivotal to the migration narratives of my interlocutors. Located in the southwest of France, the Lot is rural and inland; it was and remains one of the most sparsely inhabited and deprived departments in France, a consequence of the wider pattern of rural depopulation wrought through substantial changes in the economic base of the French national economy. In contrast to the valorisation of this space by British and other Northern European incomers, within France areas such as the Lot are viewed as marginal, (stagnant) backwaters that run counter to the French national project of progress. Perhaps then it is unsurprising that in these rural areas where agriculture is in decline, that the majority of incomers are not the non-European migrant worker populations—who earlier might have featured prominently in the statistics, fulfilling crucial roles as agricultural labourers—but are mostly Europeans (INSEE 2005; see also Buller and Hoggart 1994), taking advantage of the opportunities for buying up vacant properties. Dodd (2007) stresses the way in which these European incomers are partially 'offsetting' rural depopulation, and while we should not take such claims on face value, they nonetheless preface usefully my thinking here about the way in which Britain understands its relationship with Europe.

The ethnographic project on which this chapter builds focused on understanding the everyday lives of the British residents of the Lot and took place over 12 months between 2003 and 2005 (previously published in my monograph *The British in Rural France* (Benson 2011) and various journal articles). The project included the collation of life and migration histories, and semi-structured interviews with 49 Britons living full-time in the Lot—the length of their residence varying from those with 15 years of tenure, through to more recent incomers arriving in the last six months—as well as extensive participant observation of their and others' daily lives that took me into private homes, everyday activities and practices.

I accessed my research population through snowball sampling—as is common in ethnographic research—one person passing me onto the next. When I felt that the sample was becoming too homogeneous, I worked on ways of extending this to capture the extent of the diversity within this population. This resulted in the recruitment of interlocutors beyond those who might be considered as retirement migrants, to include those with small children as well as younger migrants who had set up small businesses in the Lot. Notably, those who took part in the research were exclusively white British, and predominantly English. Further, through their employment prior to migration all of my interlocutors would be positioned within the British middle class; however, as I discuss in detail below, their positions within this varied as a consequence of their different routes into and through the middle classes.

Migration to rural France was commonly explained through recourse to 'quality of life'; the primacy that they attributed to a better way of life framed not around work and employment but rather around lifestyle are the grounds on which I conceptualise British migration to rural France as a form of lifestyle migration (Benson and O'Reilly 2009, 2016; Hoey 2005; Benson and Osbaldiston 2016).

(Relative) Privilege and the search for a better way of life

When Karen O'Reilly and I first developed our conceptual and theoretical work on lifestyle migration it was intended as an intervention that sought to explain the migration of the *relatively affluent* in search of a better way of life (Benson and O'Reilly 2009), those whose migrations fall outside the predominant framing of migration at its intersections with labour and migration governance (Benson and O'Reilly 2016). It also sought to underpin lifestyle migration, a phrase that was gaining currency within research on privileged migration and counterurbanisation (see for example Knowles and Harper 2009; Hoey 2005), with the sociological thinking that might inform its future development as a concept.

In particular, we made a clear statement about how lifestyle, intended in its various sociological conceptualisations (e.g. as tied to consumption, as the grounds for identity-making), might be a useful framework for making sense of the migration of these *relatively affluent* and *relatively privileged* migrants (see also Benson 2015; Benson and Osbaldiston 2016). While citizenship of some of the world's most powerful nation-states (see also Croucher 2012) is a constitutive feature of such privilege—or indeed, as in the case of Britons resident in rural France, European Citizenship—our concern with the relational quality of both affluence and privilege, shifted the focus from absolute statements about the privileged or elite status of these migrants to recognise these migrations as sites for the negotiation of privilege. In crossing borders, such migrants enter into new hierarchies and find themselves variously positioned within and in relation to local social structures; yet, global asymmetries predispose their privilege to be re-validated and perhaps even enhanced in these settings (Amit 2007).

While lifestyle migration offers a framework for how we might conceive of these flows, and volunteers some concepts that might be useful, it is necessary to think about how these play out in different settings. As I have discussed through the lens of my research on North Americans in Panama, privilege and affluence—as relative statements—are socially produced in complex ways that require further unpacking and are telling of wider structural and systemic conditions (Benson 2013b, 2015; Benson and O'Reilly 2018). Importantly, privilege does not assume a location on one axis of social division; it does not only describe, for example, class, gender or ethnicity, but allows for a consideration of how different social positions might intersect. It is against this background that I focus on in this chapter, not only on the need for an understanding of privilege that identifies it as a constellation—intended here as a way of illuminating the often-invisible assemblage of characteristics through which it is constituted—but also explore how it might be put to work in thinking again about my ethnographic research with the British in rural France.

British migration to rural France: a classed phenomenon?

In this section I present a class analysis of British migration to rural France, demonstrating how this is a migration trend that is at once structured by and structuring of class formation. As I demonstrate (see also Benson 2011, 2013a), a class analysis of Britons living in the Lot reveals processes of social reproduction wrought through practices of status discrimination and social distinction.

A (very) brief introduction to cultural class analysis

Cultural class analysis offers a realist approach to understanding class that pitches social class as the central axis around which social divisions are oriented (Savage 1995). Seeking to introduce an understanding of class that recognised the possibility of a relationship between structure and agency, and inspired by Bourdieu's (1984) work on the rise of a new class formation in France—the petty *bourgeoisie*—cultural class analysis seeks to demonstrate that understanding class requires attentiveness both to its structural production and cultural articulation (Savage et al. 1992, 1995). In this framing, assets and resources play a significant role in class formation, the focus on the accumulation of these and their conversion into capitals. This analytical framing depicts the British middle class as internally differentiated and better conceptualised in the plural: the middle classes.

In recent years, this model of class analysis has become increasingly prominent in British sociology (see for example Savage 2015); recent research following this approach evaluates contemporary class formation—and by extension, class structure—in Britain through the measurement and documentation of cultural and leisure practices, mapped against financial and occupational data, social networks and connections (see for example Bennett et al. 2009; Savage et al. 2013).

The British in rural France through the model of cultural class analysis

My intention in this section is to briefly sketch out a cultural class analysis of the British in rural France, identifying class formation as dynamic, taking place in and through migration and daily life. In this way, I highlight a predisposition—a taste—for migration to rural France as well as considering how the possession and accumulation of assets and resources, read through social mobility, facilitates this migration trend.

Social mobility was most marked among the population of retirees living in the Lot at the time of the research. Perhaps unsurprisingly given the gender divisions in employment for this generation, social mobility was more marked in the case of the men taking part in the research. Vic, who had grown up in a working-class family from the East End of London, had ended his working life as the manager of a small IT company. Others such as Ron had started up their own businesses, selling these upon retirement. It was notable that these socially mobile men held in common a route that took them from school to the workplace, climbing up through the ranks to more senior positions over the course of 40-year careers. Their positions within the British occupational structure indicated not only their social position as middle class—simultaneously employers and employees—it highlighted additionally their accumulation of high levels of organisational assets.

Another common trajectory was among those who had been employed in the public sector—the teachers and civil servants who had taken the opportunities for voluntary severance, often articulated as early retirement (Benson 2010a), offered to them in the 1990s against the backdrop of wider cuts to the public sector. This surprisingly common narrative is indicative of bureaucratic assets and resources that Savage et al. (1992) identify as the grounds on which some middle-class groupings coalesce. It is worth highlighting here that such assets and resources often existed in tandem with property assets accumulated in the UK, and then deployed—wholesale—to support the *outright* purchase of homes in the Lot.

However, among my interlocutors there were also those who had established middle-class origins. This portion of the population comprised men and women of all ages; they held university degrees; preceding their migrations to rural France they had careers in a range of occupations in the public and private sector that had, for some, taken them abroad to live as expatriated workers. Many of them maintained properties back in the UK as well as owning properties in the Lot. Alannah, who had taken early retirement from the civil service, had migrated to France with her husband Daniel, an architect. They both had university degrees, had lived and worked abroad—he in Fiji, she in Finland and Australia. She spoke fluent French before migration. They had bought the house with the proceeds from one of their properties in London, but had held on to another, renting it out for an income. This was a social and spatial trajectory not limited to the population of retirees, but also common among those who had moved to France earlier in their lives. For example, aged in their early 40s, Jon and Kay had given up their jobs in research and marketing to move to rural France where they had set up a *gîte* complex; both had been to university, Jon had lived in the *Savoie* previously and spoke French fluently.

Although there is insufficient space here to explore this in detail, it is clear that the social trajectories of these migrants also translated into the lives that they anticipated leading in the Lot; it was certainly the case that those who came from middle-class backgrounds were more likely to express the aspirations to become integrated into the local community—the cultural capital endowed to them through their education manifest in their ability to learn and speak French not insignificant within this. Those from more socially mobile backgrounds appeared to hold different aspirations for living in rural France; less likely to speak French, they more often sought out the company of their compatriots, moving to rural France—with their homes and gardens—presented as the crowning pinnacle of their social mobility.

This differentiation provides further nuance to arguments about the values that the middle classes place on rural living, highlighting that beyond the headlines—beautiful scenery, slow pace of life, close-knit community— migration and aspirations for life in rural France are further influenced by individual biographies and the predispositions that these inculcate. What these broadly stated differences identify is that the British residents of the Lot are not a homogenous population; there is significant diversity despite the characterisation of this population as middle class and the shared belief that rural France offers a better way of life. Indeed, vaguely stated claims equating rural France with improved quality of life are the grounds on which intra-class distinctions play out, the relational dynamics—and concomitant pro-cesses of status discrimination—of the British middle classes remade in rural France precisely through claims to the (right) knowledge of how to live in this new environment (Benson 2011, 2013a). In other words, the context of (Brit-ish) life in rural France is ripe for processes of class formation, social group-ings coalescing around different assets, resources and values.

A distinctly British middle-class migration?

As I have examined in detail elsewhere (Benson 2011, 2013a) for my inter-locutors, migration was just one step in the project of getting to a better way of life (see also Benson and O'Reilly 2009). This search for a better way of life was more than an individualised project of self-realisation; it was thoroughly enmeshed in processes of social distinction.

This is where my analysis departs from that of the cultural class analysts; identifying not the cultural practices through which social positions are defined, but also the *relational* formation of class. My argument here channels the critiques levelled at cultural class analysis by feminist scholars such as Bradley (2014), Skeggs (2004, 2015) and Imogen Tyler (2015). As they high-light, this focus on the grounds by which class positions might be identified— e.g. consumption practices, incomes, education—neglects class relations; they advocate instead defining class as a process of classification best understood as a struggle for value. Shifting focus in this way reveals the stakes of this project, and particularly the location of this struggle for value within a dis-tinctly British sphere.

For some of my interlocutors, this was articulated through the recourse to authenticity (Benson 2013a). Living 'like the locals', restoring and furnishing properties with an eye to how they would have looked were some of the ways in which claims to authenticity were made. However, it was even more common for my interlocutors to evaluate the lives of their compatriots—often friends—as not being the right or 'proper' way to live in the Lot. In daily conversation, variations on this theme included why did people continue to try and grow English lawns? Why did people not learn to speak French? Why did they import their food and drinks from Britain? Shifting their gaze beyond the Lot, they reflected on the purported lives of fellow Britons living in the Dordogne and Spain.

In extending the project of cultural class analysis to understand class for- mation at its intersections with European integration, Favell (2008), through his discussion of the young and highly educated intra-European migrants and Andreotti et al. (2014), through their examination of the highly skilled trans- nationals, argue for a burgeoning European middle class. In both cases, these populations are understood as highly mobile and cosmopolitan, precursors to full European integration and the emergence of a European social structure stratified by class. What the preceding discussion of Britons living in the Lot demonstrates is that the discourse and actions of my interlocutors makes visible the persistence of a national field of class formation, albeit a distinctly British middle-class formation, reproduced on French soil; simply, there is no evidence that they are part of the burgeoning European middle-class Favell (2008) and Andreotti et al. (2014) identify, despite being beneficiaries—at least for the time being—of Freedom of Movement and the right to reside in another European country.

Problematising the invisibility of the white British in Europe

In this final section, I consider further the structural and systemic conditions that result in the reproduction of this distinctly British middle class in rural France and reflect on the constitution of Britishness as a particular form of whiteness in this setting.

From class to racialisation in intra-European migrations

Another way into thinking about British migrations to European destinations has been through the lens of intra-European migrations. Driven by policy concerns and the distinction in contemporary European migration govern- ance between those migrants from outside the European Union and those within, this field of research has focused on understanding migrations within Europe through the lens of the project of European integration and European citizenship (Favell 2008). Underpinning these ambitions was the free move- ment of labour and goods within the borders of Europe, free movement and the right to reside and work in another European Union state held in common by all European Union citizens. Indeed, among my interlocutors in the Lot this 'right' to settlement was often held up as the reason why they could live in rural France. However, focusing on the political project of

Europe—and in assuming its value among migrant populations—runs the risk of overlooking the wider contexts and histories within which such migrations are taking place.

As several scholars have highlighted, such 'Britishness' emerges in tandem with considerable ambivalence about what it means to be European (see for example O'Reilly 2007; Benson 2011). Their European citizenship—and reliance there-upon—awkwardly positions them within a supra-national structure; simply put, so long as certain rights remain contingent on nation and place, these mobile European citizens find themselves in an intermediary position as citizens of other European states resident in another where they do not have the full rights of that state's citizens (Ackers and Dwyer 2004; O'Reilly 2007). For example, O'Reilly's *The British on the Costa del Sol* identifies the way in which this population are a marginal community in Spain, neither colonisers nor integrated into the majority population, 'they remain essentially British; they are symbols of lost Empire, of national pride, of ambivalence towards the Other ... these Britons abroad both remind Britain of its past and intimate its future' (2000, 166).

I state this ambivalence here partly to draw attention to the project of clas-sification that lies at the heart of European integration; namely, how this redraws the distinction between citizens and the 'migrant other' around Eur-opean Union borders (Anderson 2013). This can be seen in the politics of naming, where it is common for Europeans living outside their state of origin not to be labelled as migrants. Nevertheless, there are processes of racialisation at work within this project of European integration whereby some intra-Eur-opean migrations are more prone to being racialised than others. This reveals the unequal stakes of member states and their constituent citizens within the project of Europeanisation. Indeed, as scholars working with Polish, Romanian and Hungarian populations living in Western Europe have been keen to high-light, these communities have experienced considerable racism (Fox, Morosanu and Szilassy 2015), an experience exacerbated in the immediate aftermath of Britain's referendum on their membership of the European Union.

If we understand European integration as a project that embeds both class and racial formation, how might we think differently about the production of British (migrant) communities in Europe? Neither the European middle-class populations that signal the success of European integration, nor the racialised European minorities originating in the accession states, it is timely to bring out into the open the interplay of class and racial formation in the production of these British com-munities. As I discuss below, changing the conversation in this way renders visible the *constellations of privilege* that socially produce British migration within Europe, and in particular as these are articulated in and through lifestyle migration.

Whiteness, postcoloniality and British migrations to Europe

My focus here lies in bringing to the fore the social production of whiteness at play in the settlement of my interlocutors in rural France. The analysis pre-sented here takes inspiration from scholars working on the intersections of whiteness, privilege and migration (Fechter 2007; Knowles 2005; Lehmann 2014; Leonard 2010; Lundström 2014) alongside the recognition of the post-colonial inscriptions on the British countryside (Knowles 2008; Tyler 2003, 2012; Nayak 2010).

An initial *entrée* into this discussion is the assumptions that I have been presented with repeatedly in the many years I have been working with this migrant population. At academic conferences and in everyday conversation I am regularly asked—nay, told—that the reason that many British people move abroad can be explained very simply as 'white flight', leaving Britain because of a dissatisfaction with multiculturalism. On the one hand, this is a useful discourse to think with, it brings to the fore postcolonial nostalgia and the social production of whiteness that might serve well as the crux of my argument here. On the other hand, this rationalisation was rarely explicitly stated by my interlocutors; in fact, I can recall only one couple explaining their migration in these unapologetically xenophobic terms. More to the point, it was extremely common for my interlocutors to push back against characterisations such as this, mobilising within their narratives strong and derogatory statements about their compatriots—ordinarily located outside the Lot, in Spain, in the Dordogne—who did not try to integrate. Indeed, returning to the Lot to conduct research following the UK's vote to leave the European Union, the paradoxical racist 'other' who had exercised their freedom of movement and yet voted for Brexit was frequently referred to in conversation. That I rarely came across leave voters (or supporters—given that UK citizens who have lived overseas in excess of 15 years were unable to vote in the referendum), and that when I did, their rationale for supporting this position was often far more complex, demonstrates the continuation of this trope into Brexit testimonies.

As I have argued elsewhere (Benson 2011), such rhetorical devices are more telling of the ways in which these Britons want to be seen than they are accurate depictions of other British populations abroad. The actions and practices of many of my interlocutors were a stark contrast to these caricatured populations; they revealed themselves as benign francophiles, with a genuine interest in the local history, the environment, and curiosity about the lives of their (French) neighbours. While on one level it is tempting to read this in terms that pit my interlocutors as diametrically opposed to the xenophobic Other, I argue that the narratives of these francophiles are wrought through the social production of white Britishness albeit in unremarked ways. Bringing this to the surface of the analysis below, I intend to draw attention precisely to the silence and invisibility of white privilege—persistent and insidious—to raise questions about how we might witness the social production of whiteness among these populations in the way they speak of the landscape and the people within it.

Reading whiteness in the landscape

Within *post hoc* explanations, the landscape of the Lot—its sweeping vistas, dramatic limestone cliffs—was frequently identified as a significant motivation for moving to this part of rural France (see Benson 2010b, 2011). Here I want to turn attention back on to these representations to consider these as sites for the social production of white Britishness.

While initially presented as a beautiful view, my interlocutors directing my gaze to the tableaux framed by their windows, it became clear that these landscapes held deeper symbolic and moral significance that made clear associations between this landscape and the people who lived there. As Simon, a man in his forties who regularly commuted back to London for work explained to me over our *plat du jour* one lunchtime in my local café bar, '[S]outhern French culture, it's like stepping back into an England that you do remember as a child; it's very much like 50s/60s England, our social and cultural norms ... France represents something we've lost'. Such nostalgia was a common refrain that extended into discussion of the more relaxed pace of life and was often paired with the romantic depiction of a close knit local community with a simpler way of living. At first glance, rural France appears as the paradigmatic rural idyll for these British incomers (Buller and Hoggart 1994; Barou and Prado 1995; Benson 2011). However, as other scholars have identified (see for example Agyeman and Spooner 1997; Neal 2002; Cloke 2006), such depictions when focused on the British countryside are notable for their whiteness. Katherine Tyler (2003, 2012) persuasively argues that the English countrysides are postcolonial landscapes, constituted through class and racial formations. The neglect of colonial legacies 'at home' (ibid 2012) persists into the production of what Nayak has evocatively labelled as 'the silent cartography of whiteness' (2010, 2375), whiteness invisible and unremarked, insidiously and silently reproduced.

Importantly, what becomes clear is that at the heart of how these British migrants read the landscape is an unquestioned 'right' to access living in rural France. Indeed, in returning to the Lot as part of my ongoing research on what Brexit means for UK citizens living in the EU27, I have been particularly struck by such migrants' actions to maintain this 'right'—for example, by applying for dual nationality.[2] Undoubtedly, other ethnic minority populations living in France and elsewhere would not feel so comfortable claiming this right.

In what proceeds I develop further an understanding of how class and racial formations might interplay and what this might reveal about the production of colonial legacies in British migration to rural France.

Claiming integration and the conceit of belonging

As most of my interlocutors explained, the promise of local community was part of their inspiration for moving to rural France; they yearned for opportunities to participate in local life and were disappointed when this was difficult to realise, when those opportunities were not as forthcoming as they had hoped. Importantly, this ambition is more than just rhetoric; it is something that many of my interlocutors actively work towards, pushing back against stereotypes of the British abroad that present them as not integrating and forming expatriate enclaves. As they proudly recalled the latest invitation to local events—the local hunt dinner, the party in the town hall—or more private functions at the homes of the French neighbours, how they were learning about local history through walking in the local area with the locals, they celebrated their achievements in integrating. In many ways, this valuation of the local echoes Savage et al.'s (2005) discussion of what happen to middle-class identities under globalisation; as they stress, belonging is claimed precisely through the choice to live in a particular location and the narration of this choice through biographies that demonstrate the 'fit' to this new place of residence. Indeed, read through this frame, the desire for and pursuit of the local is an almost textbook case of what Savage et al. (2005) label 'elective belonging'.

Returning to Simon's rationalisation of what rural France offers goes some way towards understanding how such claims to 'elective belonging' are wrought not only through class formation but also racial formation. Another lens onto this might be found precisely in thinking again about what claims to integration might reveal about the articulation of privilege. With this in mind, I turn to another example, this time focused on a particular location, a small village up on the limestone plateau, a remote location that had witnessed a rapid depopulation as more and more local services were removed.

Once the local primary school had been shut young families could no longer continue to live in the area and as the remaining resident population became more elderly, there were fears that their traditions and their mere existence might be forgotten. Britons who lived in the area described the village as 'dying out'; in this village of just over 90 residents, the British population constituted 10 per cent of the population. A shared joke between these Britons and their French neighbours was that the incoming British residents had significantly reduced the average age of the population, even though most of the British incomers were aged in their fifties and sixties. Keen to contribute to the area, to learn about local histories and ways of life, the Britons in this village participated in a shared project of village and community life and claimed the right to define themselves as locals.

Such claims to local belonging by the Britons in this village, just as in the case of their compatriots living in other parts of the Lot, sit alongside critiques of the failure of Others to integrate. The framing of the discussion of their compatriots was often framed around the visibility of these populations, those in the Dordogne and Spain featuring prominently; they regularly drew on a range of stereotypes commonly deployed to speak about expatriate populations—loudly shouting in English to make oneself understood, only socialising with other Britons (or English-speaking populations), not appreciating the locale and making 'little England' on foreign soil. The contrast between their willingness to take part in local life and these mythologised populations was marked.

Speaking of their compatriots in the Dordogne, one of the departments that bordered the Lot, my interlocutors made clear the stratification of the

British middle classes as it mapped onto rural France. They regularly adopted recognisable stereotypes to describe this population describing the department through the moniker 'Dordogneshire', stressing that the cricket clubs, the Conservative club as signs that this population were more interested in the creation of a 'Little England' than in living in rural France. The motifs they used to describe their compatriots living on the Spanish coastline similarly focused on their lack of integration and enclave living. Where the former were depicted as colonial-style expatriates, the latter were presented as tourists destructive to the natural environment and Spanish way of life (Benson 2011). As I argue here, these contrasts in the ways that they depicted these Other British populations are telling of the way that they see and position themselves. Their descriptions of the British living in Spain, eerily resemble the wider stigmatisation of the working class that Lawler (2005) argues is central to middle-class identity formation; through their descriptions of their neighbours in the Dordogne they define themselves in opposition to a social group who they see as occupying a higher social position. In speaking of these other British populations abroad, these migrants engaged in processes of class positioning; within a frame that understands the intersections of classed and racialised formations this also signals the presence of diversities within the social production of whiteness.

While in the first instance this is projected onto other Britons we can also read these as claims about what makes 'a good migrant'. The connotation here is that 'a good migrant' integrates, is invisible to the wider population— or rather, is indistinguishable. Importantly, the responsibility for becoming 'a good migrant' lies with the migrant. Substitute any ethnic minority population for their (white) British compatriots into these discussions and the discussion of integration becomes a far more problematic assertion signalling racist and xenophobic attitudes.

This is a useful heuristic for thinking about the work that these representations of others are doing. Lawler's (2005) work on the way that the middle classes depict the working class, and what this does for them might be useful here in terms of considering how these representations of other Brits abroad feature within a project of middle-class reproduction. We should be asking questions, as scholars of British migration, about which populations are represented in these stigmatised and stereotyped terms.

Although it is beyond the scope of this chapter, simply put, my call here is to consider how representing certain British populations that live abroad as vulgar, uncouth, as racist, or as failing to integrate constructs these as persons lacking in morals and value (Skeggs 2004; Skeggs and Loveday 2012; Lawler 2005; I. Tyler 2013). Such moral judgements by my interlocutors can therefore be understood as claims to their position as persons of value. Claims to local belonging may be further telling, however, of the co-constitution of class and racial formation. In what follows, I present a brief account of the way that local people feature in my some of my interlocutors' accounts of their integration into the local community.

Brushing over the diversity of this local population, what was particularly celebrated in their benevolent presentation of their neighbours were uncomplicated and simple ways of living, as closer to nature, and close kin and community relations. These were the lives that these Britons hoped—naively and romantically—to emulate. As they settled into their lives in the Lot, they strove to make their lives in this image; they consulted their neighbours about what and when to plant their vegetables, they developed their knowledge and understanding of the local flora and fauna, some even joined in with local activities and events. Further, the commonly held conceit that they were living 'like the locals' told of a lack of self-awareness both in the vanity of assuming similarity and about the fact that they had taken the choice to live the way they did while many of their French neighbours did not have this luxury. Several of the migrants praised the quality of life for the longevity of a local farming population who worked into their old age. Reading this differently, for many of the local French who had lived in the Lot their whole lives, there was no other option but to continue working. Their children and grandchildren no longer lived locally, their farms needed to make money for them to afford to live. And as the ethnographers of rural French life have so clearly described, this life was hard (see for example Bourdieu 1962; Rogers 1991), not the 'lifestyle' gardening that many of the British chose to undertake.

Such idealisation of the native French population is part of the way through which Britishness is produced among this migrant population. We might interpret this in line with Aldridge's (1995) discussion of Peter Mayle's best-selling autobiography, *A Year in Provence* (1989), stressing that such representations of the local French—described by Mayle using the pejorative term *paysans* (literal translation: peasant)—are more revealing of how the British in rural France see themselves and want to be seen by others. Developing this further, what becomes clear is that identifying the local populations as the Other is both a project of classification and racialisation. Reading these alongside claims to integration reveals the ways in which class and racial formation co-constitute, and highlights the persistence of colonial legacies in the making of middle-class subjects (Skeggs 2004).

Conclusion

This chapter builds on my previous writings to provide timely and developed insights into the composition of privilege among British lifestyle migrants resident in rural France. As I have argued, the migration and settlement of Britons in rural France is best understood as a process that embeds both class and racialised formations. Where previous research highlighted the role of this migration in the social reproduction of the British middle classes on French soil, recognising that the social production of whiteness is inextricably caught up in this process opens a window onto the systemic and structural conditions that promote the migration of (some) white Britons to Europe.

Simply put, reading these migrations through the lens of privilege as a multi-faceted construction reveals how colonial traces haunt Britons' conceptions of other European landscapes and peoples. Pushing back against negative representations of the British abroad, my interlocutors reveal their sympathies and desires; yet it is precisely in this process that they also make visible their value for integration and what it means to be a 'good migrant'. Deconstructing this position reveals a more complex structure of privilege than class analysis alone might reveal.

Lauding the beauty of the depopulated French countryside on their doorsteps and mobilising nostalgia for 'the England of 50 years ago', they provide rationalisations for their migration and settlement in rural France. In this way, they appropriate rural France on their own terms, through the wholesale transfer of the colonial legacies inscribed on the British countryside.

Where class analysis has the potential to explain more about how the British abroad remain part of a transnational community of Britons, this is necessarily a project that becomes self-referential to the point of excluding the wider structural and systemic conditions that make British migration possible, framing the lives, actions and practices of these populations. The unremarked whiteness of many British populations in Europe—unremarked precisely in consequence of the ideals of European integration and seeming lack of racialisation of the white British population vis-à-vis other European populations—paired with a focus on class formation, effectively silences analyses that might reveal how colonial legacies persist in the way that Britain and Britons view Europe.

The time is long overdue for a more systematic postcolonial analysis of British lifestyle migration to Europe. At its core, such an analysis should take seriously the emplacement of colonial legacies by British subjects onto European landscapes and peoples. This approach opens the space for a conversation about the (dis)continuities in British migrations the world over (Fechter and Walsh 2010). As I argue, thinking with the idea of privilege as a constellation shifts the focus from absolute understandings towards the recognition that privilege is constituted through a range of characteristics.

In the case of the British resident in Europe, this is a constellation that has included up until now European Citizenship, but which also connects with class and racial formations that have longer histories and residues. Questioning the constitution of this privilege allows for the interrogation of the extent to which British migration to Europe rests on the project of European integration and how privilege will be reshaped in the aftermath of Brexit. While up until now European citizenship has acted as a veritable dog star, the death of this star brought on by Britain's exit from the European Union might allow for the other stars to shine more brightly, for our analytic gaze to be refocused on the complexity of the *constellation of privileges* that drives British migration to Europe.

Notes

1 On 23 June 2016, the outcome of the United Kingdom's referendum on their continued membership of the European Union came out marginally in favour of leaving the European Union. In March 2017, Article 50, the legal instrument designed to allow member states to exit the Union, was triggered and negotiations between the United Kingdom and the European Union commenced concerning the terms of this 'Brexit'. A primary issue within these negotiations was citizens' rights enjoyed by an estimated 3 million non-British EU nationals living in the UK and the estimated 1.2 million UK Citizens living in the EU-27. Such rights included the freedom of movement to live and work in another European Union member state and the acquired rights (e.g. access to healthcare, the labour market) that accompany this. For UK citizens in the EU-27, as much as for non-British EU nationals living in the UK, at this point it remains unclear as to what Brexit will mean for their continued residence and employment in the places they have made their homes (Benson and Sigona 2017).

2 This project is funded through the ESRC's UK in a Changing Europe Initiative (Grant Number ES/R000875/1). To find out more visit https://brexitbritsabroad.com.

References

Ackers, L. & Dwyer, P. 2004. "Fixed laws, fluid lives: the citizenship status of post-retirement migrants in the European Union." *Ageing and Society*, 24(3): 451–475.

Agyeman, J. & Spooner, R. 1997. "Ethnicity and the rural environment." In *Contested Countryside Cultures: Rurality and Socio-cultural Marginalisation* edited by P. Cloke and J. Little, 197–210. London: Routledge.

Aldridge, A. 1995. "The English as they see others: England revealed in Provence." *The Sociological Review*, 43(3): 415–434.

Amit, V. 2007. "Structures and dispositions of travel and movement." In *Going First Class? New Approaches to Privileged Travel and Movement* edited by V. Amit, 1–14. Oxford: Berghahn Books.

Anderson, B. 2013. *Us and Them? The Dangerous Politics of Immigration Control*. Oxford: Oxford University Press.

Andreotti, A., Le Galès, P. & Moreno-Fuentes, F.J. 2014. *Globalised Minds, Roots in the City: Urban Upper-Middle Classes in Europe*. London: Wiley Blackwell.

Barou, J. & Prado, P. 1995. *Les Anglais dans nos campagnes*. Paris: L'Harmattan.

Bennett, T., Savage, M., Silva, E., Warde, A., Gayo-Cal, M. & Wright, D. 2009. *Class, Culture, Distinction*. London: Routledge.

Benson, M. 2010a. "The context and trajectory of lifestyle migration: the case of the British residents of southwest France." *European Societies*, 12(1): 45–64.

Benson, M. 2010b. "Landscape, imagination and experience: processes of emplacement among the British in rural France." *The Sociological Review*, 58(S2): 61–77.

Benson, M. 2011. *The British in Rural France: Lifestyle Migration and the Ongoing Quest for a Better Way of Life*. Manchester: Manchester University Press.

Benson, M. 2013a. "Living the 'real' dream in La France Profonde: lifestyle migration, social distinction, and the authenticities of everyday life." *Anthropological Quarterly*, 86(2): 501–525.

Benson, M. 2013b. "Postcoloniality and privilege in new lifestyle flows: the case of North Americans in Panama." *Mobilities*, 8(3): 313–330.

Benson, M. 2015. "Lifestyle migration: from the state of the art to the future of the field." *Two Homelands*, 42: 9–23.

Benson, M. & O'Reilly, K. 2009. "Migration and the search for a better way of life: a critical exploration of lifestyle migration." *The Sociological Review*, 57(4): 608–625.

Benson, M. & O'Reilly, K. 2016. "From lifestyle migration to lifestyle in migration: categories, concepts and ways of thinking." *Migration Studies*, 4(1): 20–37.

Benson, M. & O'Reilly, K. 2018. *Lifestyle Migration and Colonial Traces in Malaysia and Panama*. London: Springer.

Benson, M. & Osbaldiston, N. 2016. "Toward a critical sociology of lifestyle migration: reconceptualising migration and the search for a better way of life." *The Sociological Review*, 64(3): 407–423.

Benson, M. and Sigona, N. 2017. *Citizens' Rights*. London: UK in a Changing Europe. [Online] Available at: http://ukandeu.ac.uk/explainers/citizens-rights/.

Bourdieu, P. 1962. "Célibat et condition paysanne." *Etudes Rurales*, 5–6: 32–135.

Bourdieu, P. 1984. *Distinction: A Social Critique of the Judgement of Taste*. Harvard: Harvard University Press.

Bradley, H. 2014. "Class descriptors or class relations? Thoughts towards a critique of Savage et al." *Sociology*, 48(3): 429–436.

Buller, H. & Hoggart, K. 1994. *International Counterurbanization: British Migrants in Rural France*. Aldershot: Ashgate.

Cloke, P. 2006. "Rurality and racialized others: out of place in the countryside." In *Handbook of Rural Studies* edited by P. Cloke, T. Marsden & P. Mooney, 379–387. London: Sage.

Coles, A. & Walsh, K. 2010. "From 'Trucial state' to 'postcolonial' city? The imaginative geographies of British expatriates in Dubai." *Journal of Ethnic and Migration Studies*, 36(8): 1317–1333.

Croucher, S. 2012. "Privileged mobility in an age of globality." *Societies*, 2(1): 1–13.

Dodd, J. 2007. *The Rough Guide to the Dordogne and the Lot*. London: Rough Guides.

Favell, A. 2008. *Eurostars and Eurocities: Free Movement and Mobility in an Integrating Europe*. Chichester: Wiley Blackwell.

Fechter, A.-M. 2007. *Transnational Lives: Expatriates in Indonesia*. Farnham: Ashgate.

Fechter, A.-M. & Walsh, K. 2010. "Examining 'expatriate' continuities: postcolonial approaches to mobile professionals." *Journal of Ethnic and Migration Studies*, 36(8): 1197–1210.

Fox, J., Morosanu, L. & Szilassy, E. 2015. "Denying discrimination: status, 'race', and the whitening of Britain's new Europeans." *Journal of Ethnic and Migration Studies*, 41(5): 729–748.

Hoey, B. 2005. "From Pi to Pie: moral narratives of noneconomic migration and starting over in the postindustrial Midwest." *Journal of Contemporary Ethnography*, 34(5): 586–624.

INSEE. 2005. *Les immigrés en France*. Paris: INSEE.

Knowles, C. 2005. "Making whiteness: British lifestyle migrants in Hong Kong." In *Making Race Matter: Bodies, Space and Identity* edited by C. Alexander & C. Knowles, 90–110. Basingstoke: Palgrave Macmillan.

Knowles, C. 2008. "The landscape of post-imperial whiteness in rural Britain." *Ethnic and Racial Studies*, 31(1): 167–184.

Knowles, C. & Harper, D. 2009. *Hong Kong: Migrant Lives, Landscapes, and Journeys*. Chicago: University of Chicago Press.

Lawler, S. 2005. "Disgusted subjects: the making of middle-class identities." *The Sociological Review*, 53(3): 429–446.

Lehmann, A. 2014. *Transnational Lives in China: Expatriates in a Globalizing City.* Basingstoke: Palgrave Macmillan.

Leonard, P. 2010. *Expatriate Identities in Postcolonial Organizations: Working Whiteness.* Farnham: Ashgate.

Lundström, C. 2014. *White Migrations: Gender, Whiteness and Privilege in Transnational Migration.* Basingstoke: Palgrave.

Mayle, P. 1989. *A Year in Provence.* London: Pan Books.

Nayak, A. 2010. "Race, affect, and emotion: young people, racism and graffiti in the postcolonial English suburb." *Environment and Planning A*, 42(10): 2370–2392.

Neal, S. 2002. "Rural landscape, representation and racism: examining multicultural citizenship and policy-making in the English countryside." *Ethnic and Racial Studies*, 25(3): 442–461.

O'Reilly, K. 2000. *The British on the Costa del Sol.* London: Routledge.

O'Reilly, K. 2007. "Intra-European migration and the mobility—enclosure dialectic." *Sociology*, 41(2): 277–293.

Rogers, S. 1991. *Shaping Modern Times in Rural France.* Princeton, NJ: Princeton University Press.

Savage, M. 1995. "Class analysis and social research." In *Social Change and the Middle Classes* edited by M. Savage & T. Butler, 15–25. London: UCL Press Limited.

Savage, M. 2015. *Social Class in the 21st Century.* London: Penguin.

Savage, M., Bagnall, G. & Longhurst, B. 2005. *Globalization and Belonging.* London: Sage.

Savage, M., Barlow, J. & Dickens, P. 1992. *Property, Bureaucracy and Culture: Middle Class Formation in Contemporary Britain.* London: Routledge.

Savage, M., Devine, F., Cunningham, N., Taylor, M., Li Yaojun, Hjellbrekke, J., Le Roux, B., Friedman, S. & Miles, A. 2013. "A new model of social class? Findings from the BBC's Great British Class Survey experiment." *Sociology*, 4(2): 219–250.

Skeggs, B. 2004. *Class, Self, Culture.* London: Routledge.

Skeggs, B. 2015. "Introduction: stratification or exploitation, domination, dispossession and devaluation?" *The Sociological Review*, 63(2): 205–222.

Skeggs, B. & Loveday, V. 2012. "Struggles for value: value practices, injustice, judgment, affect and the idea of class." *The British Journal of Sociology*, 63(3): 472–490.

Tyler, I. 2013. *Revolting Subjects: Social Abjection and Resistance in Neoliberal Britain.* London: Zed Books.

Tyler, I. 2015. "Classificatory struggles: class, culture and inequality in neoliberal times." *The Sociological Review*, 63(2): 493–511.

Tyler, K. 2012. *Whiteness, Class and the Legacies of Empire: On Home Ground.* Basingstoke: Palgrave Macmillan.

Tyler, K. 2003. "The racialised and classed constitution of English village life." *Ethnos*, 68(3): 391–412.

3   Home consumption and belonging among British migrants in Western Australia

Gillian Abel

Introduction

This chapter explores the home consumption practices of a sample of British migrant women, mainly nurses and midwives, living in Perth, Western Australia. The structural underpinning of their migration was a continuing demand for skilled labour in the Australian health sector, which facilitated their arrival during the period 2000 to 2005. I argue, however, that there is a gap between the state category of skilled migrant, into which these women must fit, and the subjective reasons for their migration, most often reported as a lifestyle choice (see also Forrest, Johnston and Poulsen 2014). This chapter, looks beyond the bureaucratic classification of skilled migrant to the lived experience of these women in a country where 'acquisition of the family home is likened to the attainment of the national dream' (Andrews and Caldera Sánchez 2011, p.208). In particular, this chapter focuses on the act of house buying as a significant factor in their efforts to establish belonging.

In this migration context, buying the right house is central to buying the right lifestyle and therefore a key act in what might be described as a consumption-led migration (King 2002). In both the Australian and British setting there is a 'normative preference for homeownership' where it is viewed as a 'reflection of having achieved a desirable social status' (Easthorpe 2014, p.591). In a similar vein, it is argued that migrant homeownership demonstrates 'integration into Australia's dominantly middle-class society' (Forrest, et al. 2014, p.109). Earlier British migrants firmly embraced this fundamental of the lifestyle package commonly referred to as the 'Australian Dream' (Hammerton & Thomson 2005), as have the more recent cohort examined here.

In foregrounding homeownership in this chapter, I acknowledge the complexity associated with the concept of home, said to comprise 'a place/site, a set of feelings/cultural meanings, and the relations between the two' (Blunt and Dowling 2006, p.2). This relational conception of home comes to prominence in the context of migration. For Jacobs, the home is:

[T]he pivotal point from which one reorientates oneself, not simply to one's new neighbors, new nation, and new society, but also to one's old home, one's memories, one's responsibilities to family left behind or moved on elsewhere.

(Jacobs 2006, p.8)

While for Walsh, 'the negotiative effort that goes into making home meaningful, as well as the salience of material culture in this process, are highlighted by international migration' (2011, p.516).

Out-migration from the United Kingdom (UK) is recognised, not least in the conception of this volume, as somewhat under-researched (see also Lunt 2008; Sriskandarajah & Drew 2006). For King (2002, p.102), migrations, 'can be spectacular or mundane […] regarded as problematic or non-problematic' and British migration to Australia arguably fits the latter categories. While immigration is a favourite topic of both the British and Australian popular press, certain aspects remain under the radar. The designation of British migrants in Australia as 'invisible' (Hammerton & Thomson 2005, p.8) is particularly pertinent when viewed relative to the 'the composite folk devil identity of the Muslim-terrorist-refugee' (Martin 2015, p.308) which features so strongly in the immigration discourse of both countries. As King observes 'the nature of the "spectacle" is often exaggerated and distorted by its media portrayal and politicisation' (2002, p.102). This chapter offers a counterbalance to such polemical discourse in its attention to an arguably less visible migrant cohort.

The focus of my doctoral thesis, on which this chapter draws, was the settlement experience of British women, mainly nurses. Previous studies have shown that common stereotypes of mobility focus on either disadvantaged labour migrants or, at the other extreme, on highly mobile elites who move globally with apparent ease (Conradson & Latham 2005a; Conradson & Latham 2005b; Favell, Feldblum & Smith 2007; Scott 2006). In such hierarchical understandings of international migration, it can be argued that nurses occupy what Conradson and Latham (2005b) call a 'middling' space. Mar (2005, p.367) concurs, referring to 'middling migrants' as 'mostly professionals and para-professionals such as nurses, engineers and social workers […] not "high flying" business migrants'. They differ for example from the Working Holiday Makers (WHMs) featured in Clarke's research (2004; 2005) and from those who travel at company expense (see, for example, Beaverstock 2005).

In addition to such 'middling' status various other factors combine to hide the nuances of these women's migration experiences. Their gender is one significant factor, in that the movement of skilled female migrants, either as primary visa holder or accompanying person, remains under researched (Kofman 2012). In addition, presumptions of ethnic similitude with the host population assume a supposedly easy fit, due to common language and historical connections. A further contributing factor is their positioning in 'the middle class of suburbia', which Miller (2010, p.84) referred to as 'a class so universally derided by theorists that no one had bothered to theorize them properly'. While I argue that the factors listed above contribute to the invisibility of these migrants, they are simultaneously determining factors in their consumption patterns. For Wiles, middling migrants are 'not motivated to move by narrowly economic or political reasons [...rather,] motivations for moving tend to be social and cultural, for opportunities for travel and pleasure' (2008, p.117). As white women from the United Kingdom, these women have desires and consumption patterns in common with those of similar gender, class, life-stage and ethnic or racialised identity.

Methodology

Recruitment efforts began in late 2006 with a traditional snowballing method. I also utilised Web Based Discussion Forums (WBDFs) popular with the British migrant community in Western Australia and placed posters in the major hospitals in the Perth metropolitan region. A further, and by far the most successful action, was an advert placed in 'On Board', a twice-yearly newsletter issued by the Nurses Board of WA: this created a significant response. Forty-seven women were subsequently interviewed in Western Australia. Thirty-three were nurses, eight were working as midwives, and one was a recruitment agent, specialising in the placement of nurses. I also interviewed five non-nurses whom I met at a weekly lunchtime gathering of British women, promoted on the WBDFs. I was introduced to this group by a nurse I interviewed and I attended regularly, over approximately one year, from early 2007. Unlike the nurses in the study, most of the women who attended the lunches were not the primary visa holder; rather their husbands' skills, commonly in the construction trades, and therefore also positioning them as 'middling', facilitated their migration. The discussion in this chapter focuses mainly on the nurses and midwives interviewed but it would be remiss not to acknowledge the influence of those who contributed to the broader doctoral project.

While a shared language and history have undoubtedly benefited British migrants, there are still formalities to be addressed in negotiating Australia's notably 'exclusive-protectionist' migration programme (Iredale 2005, p.161). Australia has long sought immigrants as a labour force, however, since the late 1980s there has been an increased focus on the management of immigration flows. Immigration in Australia is managed in three streams, skilled, family and humanitarian, with an ever-increasing emphasis on the skilled stream (Phillips & Spinks 2012). The participants in this study entered through the skilled stream in either the general skilled migration or employer nomination categories. Both categories are assessed using a general skilled migration points test which stipulates, among others, various criteria such as English language competency, age at time of application, and skills or qualifications recognised by an approved body within Australia.[1]

The women in the study were, in the main, aged between 30 to 50 years and most had school age children still living at home. Most interviews were carried out in the women's homes, some took place in coffee shops at the participant's request. The ethnographic interview, as described by Baldassar, Baldock and Wilding (2006), takes into consideration geographical distance between participants but also allows for a degree of participation in their day-to-day, for example, through visiting their homes and attending social functions with them. This recognises that what is traditionally known as participant observation is often challenging in the urban/suburban research situation. The women's stories[2] and discussion of the WBDFs are augmented by discussion of the *Whingeing Pom*, a locally produced and distributed magazine aimed at British migrants, launched during the research period but which has since ceased publication. The bulk of the data analysis was carried out between 2009 and 2010.

During the semi-structured interviews, I asked whether the women had experienced any watershed moments in the settling process and it was this prompt, which often elicited a response regarding home ownership. Even for those who had not yet purchased a home in Australia it was generally declared as an intention and was strongly associated with the achievement of a better lifestyle. This corresponds with my own settlement experience and those of family and friends who have similar migration stories. My personal experience is clearly central to the design, execution and analysis of this research. As I established a new life here in Western Australia, I had a growing recognition that neither I, nor my fellow Brits, were viewed as 'migrants' in the same sense as many other nationalities. Such experience generated initial research questions around migration hierarchies and practices of inclusion and exclusion in the migration discourse; to paraphrase, all economic migrants are not created equal. Such 'insider' status as a researcher, whether embraced or denied, comes with the potential for 'particular kinds of insights and oversights' Voloder (2008, p.30). Such status can offer a shared point of reference such as nationality or gender and yet simultaneously make differences such as socio-economic status or generation significantly more obvious than they might have been where no assumptions of similitude existed (Ganga & Scott 2006). Ethnographic endeavours have been described as 'inherently partial – committed and incomplete' (Clifford 1986, p.7), this work is no different. However, as noted in the introduction, there is some discrepancy between the bureaucratic categorisation of these migrants and their lived experience, which an ethnographic approach can go some way to address.

In the market for a better lifestyle: theorising consumption led migration

The main protagonists in this chapter are a group of arguably privileged, white, English speaking women who, in general, came to Australia for a 'better life' or in many cases a better life for their children. Capitalising on their skills, these women moved approximately 9,000 miles with expectations of a British summer-holiday-style destination which offered things including sun, beaches and a slower pace of life. The women in this study were, as King suggests:

> [A]ble quite explicitly to 'shop' for opportunities and destinations, measuring the costs and benefits of risk, insecurity, quality of life, anticipated income, cultural (un)familiarity, and existence of social kin and contacts.
>
> (King 2002, p.95)

Importantly for this chapter, the move offered the ability to consume in ways which life and opportunity structures in Britain did not allow and which became central to how they imagined their migrant trajectories.

Australia has long been heavily promoted to British citizens as a desirable destination by numerous sources including governments, public and private sector employers, tourist boards, television companies and, importantly, by other migrants. Hammerton and Thomson (2005, p.40) note that the late 1950s saw Australia 'sold as a country where young families would prosper in a modern society with familiar British characteristics but a superior climate and lifestyle'. More contemporary recruitment efforts echo such sentiments; one such advert from the Western Australian government seeking health workers from the United Kingdom, showed a picture of a couple snorkelling in turquoise water and claimed that 'Western Australians will tell you that our state offers the best lifestyle on the globe' (reproduced in Abel 2016, p.14). The sales pitch is perpetuated and continues to permeate the British imaginary.

Theorising such mobility involves a departure from traditional areas of focus in migration studies, such as low-skilled high volume movements, hence the emergence of lifestyle migration as a significant area of scholarship. Lifestyle migration has been defined as the movement of 'relatively affluent individuals, moving either part-time or full-time, permanently or temporarily, to places which, for various reasons, signify for the migrants something loosely defined as quality of life' (Benson & O'Reilly 2009b, p.621). This ties in with my earlier argument that the 'middling' British migrants who are the subject of this chapter are under-researched, and while the lifestyle migration category may not be an exact fit, there is significant crossover in the migrant experiences to make the literature an important resource in discussing this contemporary British migration to Australia.

One of the most obvious points of similarity is that movement included in the lifestyle migration bracket is generally regarded as elective. As Benson and Osbaldiston report:

> Rather than a focus on production and the involuntary nature of many migrations, lifestyle migration appears to be driven by consumption and is optional and voluntary, privileging cultural motifs of destinations and mobilities (2014, p.3).

Importantly, in relation to the notion that this is a migration of choice, the movement of British migrants to Australia in the post-WWII period has been described as being 'more likely to be stimulated by a sense of heightened expectations than desperation to escape austerity' (Hammerton & Thomson 2005, p.1). For Appleyard (1964, p.213) the migration was an 'opportunity to better their socio-economic conditions rather than change their socio-economic status'. Hollinsworth, similarly stated that '[m]ost British migrants came for a better life for their children, the climate and housing standards rather than direct economic advancement' (1998, p.233). I refer to these older examples to illustrate continuity present in narratives of British migration to Australia.

When discussing such voluntary migration, it is important to acknowledge the various influences on migrant decision-making. While the migrants in my study claim to be searching for a better lifestyle this tends to be an aspiration commensurate with their existing life. As Benson and O'Reilly note of those they term lifestyle migrants:

> They bring with them skills, expectations, and aspirations from their lives before migration. Their lifestyle choices thus remain mediated by their habitus, and framed by their levels of symbolic capital. In other words, their relative symbolic capital (incorporating educational, cultural, and social capital) impacts on the decision to migrate and the destinations chosen, but also the life then led in the destination.
>
> (Benson & O'Reilly 2009b, p.618)

Arguably then, their lives continue in many ways as before, albeit in a more spacious home in the sunshine, often with a swimming pool in the backyard.

Location, location, location: the popularity of Perth's coastal suburbs

'Location, location, location' is a heavily used cliché that claims that the three most important things about a property can be reduced to its location. Notably, since 2000, it has been the title of a popular British lifestyle television series concerning the purchase of residential property that, in 2012, resulted in an Australian-centred spin-off series. Here, I use the phrase to highlight the importance of location in the migrant decision-making for those in my study, and to highlight the significance of conducting the research in the Perth Metropolitan area.

A report published around the time my research was carried out estimated that 15 per cent of Britons who live permanently abroad are in Australia; only the United States of America, at around 20 per cent, hosts more (Sriskandarajah & Drew 2006, p.119). Australia was also cited as the destination of preference for Britons seriously considering moving abroad; the figure standing at around 22 per cent compared with only 12 per cent for the next most popular choice, Canada (Sriskandarajah & Drew 2006, p.121). Australian population statistics tell a similar story about its popularity as a destination for British migrants. The *Characteristics of Recent Migrants* (CoRMS) report (Australian Bureau Statistics (ABS) 2011), which concerns those who entered the country between 2001 and 2010, the period during which my fieldwork was carried out, indicated that the majority of recent migrants were born in the United Kingdom[3]. Further, the report notes that most recent migrants from the United Kingdom chose to settle in Western Australia (34 per cent) followed by South Australia (21 per cent). Despite the statistical significance of this migrant group they remain under-represented in the migration literature.

Perth is then an eminently suitable location for an investigation of British migrants. In addition to the statistics on recent migrants quoted above, Western Australia was recognised as the state with the highest percentage of United Kingdom born residents, with the figure standing at around 12 per cent of the Western Australian population in the 2011 Australian Bureau of Statistics Census of Population and Housing. This data also identifies what Salt (2012) has called Australia's 'five most "British" suburbs' namely: Jindalee (43 per cent), Mindarie (34 per cent), Connolly (33 per cent), Burns Beach (32 per cent) and Carramar (32 per cent), all of which are located north of the Central Business District (CBD) in Perth's western coastal corridor, with Carramar, at approximately 8 kilometres from the coast, being the furthest inland. Life in the suburbs has been an identifiable trend amongst British migrants in Perth since the post-war period (Hammerton & Colborne 2001; Hammerton & Thomson 2005) and continues to be so. The topic of location within the Perth metropolitan region, which epitomises urban sprawl, is one regularly discussed and hotly debated on the WBDF's and those I interviewed displayed definite tendencies for the Northern coastal corridor living as close to the CBD and the ocean as finances would allow. This was reflected in the statistics regarding the most British suburbs presented earlier and is in keeping with the perception of what constitutes a better life among these recent migrants.

The main reason for the concentration of British migrants in these suburbs is the attraction of the coast. The beach and ocean are two of the enduring symbols of the lifestyle package that makes Australia so attractive to those from the United Kingdom, which speaks to the influence of tourism on the migration decision-making of these migrants. 'Coastal lifestyle migration emphasises escape, leisure, relaxation and "tourism as a way of life"' (Benson and O'Reilly 2009b, p.612), and the following quote from Angela, a registered nurse, gives some sense of this:

> We used to go to Spain a lot on holiday and you always get those holiday blues going back home, and we just thought well this has been great. Eating outside every night at the apartment, the kids jumping in the pool and everything. I think on that holiday we decided to take stock of our lives, we thought well we are in our 30s now, are we going to stay where

we are for the rest of our lives or are we going to try to do something a bit different? I had a friend who at that point had been in Western Australia for 7 years and another in the process of moving over here and then we started to think, well should we think about emigrating and if so where would we go? We needed to go somewhere where they spoke English basically, so Spain, although it is a lot closer, was out of the picture. We went back home and we were craving that kind of lifestyle.

Angela's was not an isolated example and the nursing agency employee I interviewed admitted that she actively discouraged migrant nurses from working at Armadale hospital, based in an inland suburb around 35 kilometres south-east of the CBD, her opinion being that their chances of successful settlement were greater when they chose to live in the beachside suburbs, anywhere from Mandurah in the south to the likes of Hillarys, Quinns Rocks and Mindarie in the north.

This is also important in that while British migration to Australia does not follow such a definite pattern of chain migration as has been identified in other migration studies it was common for participants to note that they had been introduced to an area by friends or family or in some cases more tenuous links. Migrants in many cases follow established routes to known destinations (Hugo & Harris 2011), and the British in this study are no exception. Kate, for instance, did not personally know anyone in Western Australia on arrival but, like many of the women interviewed, had a contact through mutual friends. Kate's contact picked the family up from the airport and housed them for a couple of weeks on arrival. Kate noted that she had assumed from her internet-based research prior to moving that they would most likely settle in the Northern suburbs, but the initial connection to the southern suburb of Baldivis led them to purchase a house and land package there.

Yvonne's move with her husband and four children was informed by an exchange teacher who had spent time at the children's school in England and advised they look around the northern beachside suburb of Sorrento. Another respondent said that North of the river had always been her choice due to having British friends who lived in Mindarie, a beachside suburb around 30 kilometres north of Perth CBD. This family focused their search in the newer suburbs around that area, which have blossomed over the last decade or so. However, the changing property market in both the United Kingdom and Perth, with the former stagnating as the latter surged ahead, conspired against them and caused them to settle somewhat further north than they had initially intended. Perth property prices were reported to have doubled between 2001 and 2005[4] at a time when the Australian dollar was strengthening against the British pound. It was therefore common for migrants to have to adjust their ideas about where to live by the time they arrived in Australia. Some timed things more favourably. Helen, who arrived in 2002, realised how fortunate they were to sell their home in the United Kingdom so quickly and be in the position to purchase a home in Australia. 'We tripled our money coming over, we were, looking back just very lucky'. She acknowledged that a later arrival could have been very costly for the family.

'I don't know anyone who is happy for any length of time in a rental': the preference for homeownership

Having selected their suburb, the next step was finding a suitable home to purchase. Most of those I interviewed had been homeowners, or in the process of purchasing in the U.K. prior to their migration. This corresponds with information from the Office of National Statistics (2013) which states that in England and Wales in 2001 69 per cent of homes were owned or being purchased compared with 31 per cent rented. The importance of home ownership over living in rented accommodation is highlighted in the quote from Phyllis:

> We lived in rented accommodation, two lots of rented, [...] and then we decided to look for a house because I wasn't settling very well and I thought if I had my own home I'd settle better.

As with earlier migrants' rental is viewed as an 'interim measure before realization of the ultimate dream of home ownership' (Hammerton & Thomson 2005, p.219). Another participant said, 'well I don't know anyone who is happy for any length of time in a rental'. She had purchased her home only four days after arriving in Western Australia and said she loved the property, it was in an established area not too far from the beach and close to the school her children were to attend.

For Heather, the purchase of the family home, in Perth's northern suburbs, appeared to offer an opportunity to consolidate their migration decision:

> Now that we are in our own home I am thinking I am going to drop my hours. You know, you are a long time dead. I don't need to work full time [...] I thought even if I just dropped one day it could make a huge difference, just by being able to take the boys to school or going to meet the girls that do the walk on a Friday. So, I can be a bit more flexible, or accessible, rather than just saying 'oh no sorry I can't do that' or 'oh no sorry I can't do that either'.

Heather was optimistic their new home would become a hub for their own and their children's friends, as their home in the United Kingdom had been, but not, she implied, the rental property they had lived in on their arrival in Australia. This accords with Richard's conclusion, in her study of suburban Australia, that 'family life is entered via home ownership' (1990, p.117). For this family, like others, the purchase of their new home is a focal point in their efforts to establish themselves in Western Australia.

Elizabeth, noted that she would rather have lived closer to the city to reduce her commute but, in the time it took to process their visas and then sell their house in the United Kingdom, that desire had become financially untenable. Both Heather and Elizabeth's experiences invoke consideration of the observation that '[o]ne has the Paris that goes with one's economic capital and also with one's cultural and social capital' (Bourdieu 1983, p.128) and further with the idea expressed by Douglas and Isherwood (1978) that consumption is about availability in society, arguing that the better off are more available. As previously noted all the women I interviewed, or their husbands, were in a position where they had to work here in Australia. Indeed, this is not so different from the migrants who feature in the broader lifestyle migration literature. With the exception of self-funded retirees, many are involved in some economic activity, however, it is often reported in terms of a positive feature of their lifestyle change rather than as a necessity (O'Reilly & Benson 2009).

Size matters

New homes in Australia have been reported to measure around three to four times the size of their less spacious counterparts in the United Kingdom (Dowling and Power 2012, James 2009). This size differential was commonly recognised amongst my participants and often used as a point of comparison regarding their pre- and post-migration lives. The ability to purchase and occupy, in Australia, a significantly greater living area, including home and garden, than was possible in the United Kingdom emerges in the excerpts to follow as an important factor in the migrant narrative.

It transpired in conversation with Joyce that her family had planned to spend only two years in Australia before returning to the United Kingdom; but enjoying it as much as they did, decided to stay and purchased their own home. This was something Joyce found herself unable to tell her parents even when they came to visit; initially they were given the impression it was a rental property and eventually Joyce asked her husband to break the news, noting that she had tried to but just could not get the words out. Despite Joyce's concerns she found her parents supportive and understanding, as she explained it:

> How the hell could you buy a house like this in England? Well you couldn't, no way. It is a different world, mind you we got in just in time but … it is a different world. When you think what you'd pay for this house in England, this is beyond even the wildest dreams. You couldn't have a property like that you know, I've got fruit trees and everything. My Mum can't get over my grapefruit tree and my lemon tree.

Joyce's reasoning for the positive reaction was that her parents could fully appreciate the difference in living standards for her family between the United Kingdom and Western Australia.

Brenda, similarly used a visit by relatives from the United Kingdom to illustrate her conception of the difference in housing:

> People will think oh well but I had a three-bedroom house in the UK, but the chances are that here you'll have three bedrooms plus a games room, and a pool in the back yard! My husband's brother made me laugh, I picked them up from the airport and he walked through the door and up the hallway and said, 'which part is yours' and I went 'all of it', you could fit his house in just our lounge room. Ours [in England] was just a semi, a three-bedroom semi that we put an extension on [...] it was quite a decent area, but I mean it is not a patch on this.

In addition, Kim succinctly noted, 'it is only a small four by two, but it is ours; we've got a pool. You couldn't do that in England' ably summing up the prevailing sentiment.

The tendency to highlight positive aspects of the migration, such as the larger house size, speaks to an understanding of this elective migration as a comparative project (Benson and O'Reilly 2009b, King 2002). In a similar vein Baldassar (2001, 2007) has discussed the importance to Italian migrants of achieving what is regarded as a successful migration. To be seen to be successful in the new location justifies the move for a variety of reasons, which may include financial reasons and, as demonstrated in the case of my participant Joyce, feelings of guilt at leaving in the first instance.

The dream home narrative: as featured in the *Whingeing Pom*

As with the general Australian population, many migrants decide that the best way to get what they want in a home is to build rather than purchase an established home. The February/March 2009 issue of the *Whingeing Pom* displays the headline 'Bruce the Builder: Constructing Your Dream Home Aussie Style' alongside an image of a sandcastle topped with an Australian flag built beside the ubiquitous Australian Colorbond^{TM} fence. Inside the magazine the article begins:

> Remember that time in primary school when Mrs Butcher, or whoever, asked you to imagine your perfect home? Lots of excited discussion followed. Talk of mansions [...] whole rooms put aside for massive televisions, a swimming pool [...] Well [the article continues] clever old Mrs Butcher was preparing the way for any of her tiny charges who would one day wave goodbye to windswept Wymondham and head off around the world to set up their stall in Australia.

The article goes on to discuss the experiences of a couple recently arrived from the United Kingdom, with their two children, who chose to build their 'dream home' in Western Australia.

The couple are quoted in the article as saying 'We hadn't planned it [building] ... [w]e were looking at long-term rental or buying a house ... there were some nice places around but nothing that seemed perfect'. Regarding the house that they built, the author notes that:

> [W]hat they'd ordered ticked all the boxes – a theatre room, study, five bedrooms, two with en-suite, an extra-wide garage to fit both their cars ... a computer nook and, of course, a pool out the back ... The property is basically designed around their family, even down to the second en-suite being on the first floor [sic] negating the need for aged relatives to negotiate the stairs when they come and stay.

In addition to an emphasis on a better life for their children the consideration of family remaining in the UK when purchasing, renting or indeed, designing a home, featured in many conversations during my research. Due to the distance and cost of the journey, visits from the UK are often lengthy, commonly at least one month, and provision for this, particularly in the form of a spare bedroom was important.

Such transnational aspects of the move are further visible in advertising in the February/March 2010 edition of the *Whingeing Pom*, which contains a double page advert for Perth building company Summit Homes. Accompanying a photo of a young man and woman is the following text:

> Many builders showed them 'A Spacious Home'[.] But we knew they wanted 'A We Can't Wait To Send Back Photos Of Us In Our Huge Kitchen And Out On Our Sunny Alfresco Area Home'. At Summit Homes we spend less time talking about ourselves and more time listening to what our customers say, so we get to know exactly what is important to them. That's how we've helped so many British people into their perfect Summit home, with all the wide-open space to enjoy that great West Australian freedom.

The text in this advert is significant in that it sums up the type of desires frequently expressed by British migrants in relation to their expected lifestyle in Australia. Further, through mentioning the sending back of photos, it acknowledges the importance of communicating the success of the migration to those left behind.

A further full-page advert featured in the same magazine issue speaks to similar aspirations and expectations. It comprises a list of what another local company, Jaxon Builders, believes are 'a few of the main features you'll need to take full advantage of the great local climate and lifestyle'. The list includes an 'undercover alfresco area' the ubiquitous Australian barbecue or 'barbie' and 'two dunnies' the latter accompanied by the explanation 'With all the indoor and outdoor activity (especially when you put in your swimming pool) you're going to need at least two toilets'. This declaration of need can be related to Miller's (1998) observations regarding heightened expectations and the social necessity to consume in order to fit into particular lifestyle situations, he states:

> Whether it is the speed with which a medical operation should be carried out, or the number of toilets in a family house, modern consumption has certainly demonstrated how hard it is to overestimate our capacity to need as opposed to merely want.
>
> (Miller 1998, p.138)

One interviewee demonstrated this in revealing that she occasionally worked night shifts while her husband worked away leaving their children, aged 13 and 8 at the time, at home alone. She explained that the family needed her wages to finance building their home, the ubiquitous four by two with both pool and spa, but also revealed that they would be mortgage free here in Australia.

While one can get a sense here that the migrants seek to flaunt their economic mobility it is important to view this in context as Miller (1998) suggests. Dowling and Power (2012, p.613) have argued that the 'lens of status' is not sufficient to explain the trend towards larger homes amongst their respondents. They offer a broader consideration in which 'bigger houses are a spatial accommodation of the complexity of contemporary middle-class family life' (2012, p.607). This speaks to both concerns for family continuity as an element of home ownership (Dupuis & Thorns 1996) and further to the importance of the concept of social reproduction on these migrant stories, relating as they often do to homemaking and the notion of a better life for the children (Kofman 2012). Leaving aside arguments about status and sustainability it is not difficult to see ways in which the housing choices of those in my study reflect societal trends, for example as demonstrated in the advertisements discussed.

Taking the plunge: the backyard swimming pool as a symbol of a successful migration

In addition to the family home, one of the significant markers of a successful migration to Australia takes the form of a backyard swimming pool. This can be seen in the actions of the migrants and in the advertising aimed at them. A swimming pool becomes almost obligatory, as my husband and I discovered soon after our arrival in Perth. Upon hearing that our newly purchased home did not have a pool, a well-established local businessman, originally from Yorkshire, exclaimed incredulously, 'What the f**k's the point of coming to Australia if you don't have a swimming pool?' Whether you have a pool is one of the first questions those remaining in the United Kingdom ask of you, to the point that one friend commented after their swimming pool was installed: 'at least I won't have to lie to people back home now'. This was clearly said in jest but it speaks to expectations regarding migrant life for the British in Australia and the normative values that I have already discussed as influencing consumption. Pools, are an established symbol of a successful migration to the point that during the writing of this chapter a British migrant, introduced through a mutual friend, expressed concern that her children, born and raised in Australia, had missed out on an essential component of an 'Aussie childhood' because they never had a backyard pool.

Other similar examples emerged throughout the research, Joyce responded emphatically when asked what she took the most pleasure from in her decision to move to Western Australia:

> The weather, and the blue sky, and I mean how many houses do you know in England that've got a swimming pool? You know it is a millionaire's job, and a lot of Australians don't realise ... in England you don't have a pool unless you are very rich.

One family had just had their swimming pool installed in their back yard and were in the process of completing the patio area around it when I conducted our interview. When I asked whether they had hosted visitors from home my respondent answered only her husband's parents. She then qualified this statement by saying, with reference to their swimming pool, 'I told everyone not to come out until next year when it's all done'. Another stated that the swimming pool was a must have for the two teenage daughters who moved with her when they were looking at houses:

> The kids said we've got to have a swimming pool; we've got to have a swimming pool. So, we bought the house with the swimming pool!

She further noted the novelty value for her grandchildren back in the United Kingdom:

> It is great talking to them on the phone just now. One has just turned three and the other is soon five and says, "I can swim Nanny, without armbands". And I said oh and where are you going to swim then? "In your pool" she replied, she thinks it is great that we've got a pool in the garden.

These examples, correspond with Benson's observation that the backyard swimming pools installed by British lifestyle migrants in France, 'were an emblem of the leisure opportunities that these new surroundings offered' (2011, p.98). Further, they highlight the ever-present significance of the both the transnational, and the need to demonstrate a successful migration, through the common inclusion of family in the UK in migrant narratives concerning the backyard pool.

Conclusion

This chapter contributes to our understanding of contemporary international British mobilities by focusing on the home consumption practices of a cohort of British migrant women who otherwise occupy a position of relative invisibility in the broader discussion of both British out-migration and Australian immigration (Lunt 2008; Sriskandarajah and Drew 2006). As 'middling' migrants their ability to shop for opportunities is, in part, what makes these women invisible. While their invisibility is compounded by their particular combination of gender, class, life stage and ethnic or racialised identity, this can be mitigated by recognising that these same characteristics are determining factors in their consumption patterns (Miller 2010), and investigating such, as I have done in relation to home ownership.

In certain aspects of their lives, and commonly stemming from the shared British Australian history, these women enjoy privileges not afforded all Australian immigrants. In this regard, lifestyle migration scholarship (Benson & O'Reilly 2009a, 2009b) and its departure from the traditional areas of focus in migration studies proves useful in illuminating these migrant experiences. Further, in heeding Miller's (2010) advice not to ignore the 'middle class of suburbia' I have challenged the invisibility conferred by such privilege. While it would be simple to dismiss the purchase of large coastal homes with backyard swimming pools as conspicuous consumption, or as flaunting one's economic mobility, a closer look demonstrates how these purchases are enmeshed in an intricate web of societal expectations and social reproduction. As Dowling and Power (2012) highlight, the complexity inherent in middle-class homes and lives cannot be appreciated by a focus on status alone.

I have argued that the home, conceptualised relationally (Blunt & Dowling 2006), is a central feature in the transnational comparative project which constitutes this migration and, along with the backyard swimming pool, makes a significant contribution to facilitating belonging. Buying into the lifestyle is key to belonging among this cohort of Western Australia's British migrant community.

Notes

1 Requirements have changed somewhat since the data in this study was collected with a new points test introduced in 2011.
2 Pseudonyms have been used and certain characteristics, for example, occupations and place names, have been changed to protect participants' confidentiality.
3 Later ABS data from 2010–2011 showed a significant change, for the first time the United Kingdom was not the largest source of migrants with China topping the list as the source of 14 per cent of the migrant population above both the United Kingdom and India on 10 per cent each.
4 http://reiwa.com.au/uploadedfiles/public/content/the_wa_market/house-prices-2013-web.pdf.

References

Abel, G. 2016. *A Better Life [Style]: British Migration to Western Australia Made Visible through the Lens of Consumption*. PhD Thesis. University of Western Australia, Perth WA.

Appleyard, R.T. 1964. *British Emigration to Australia*. Canberra: The Australian National University Canberra.

Andrews, D. & Caldera Sánchez, A. 2011. 'The Evolution of Homeownership Rates in Selected OECD Countries: Demographic and Public Policy Influences'. *OECD Journal: Economic Studies*, 1: 207–243.

Australian Bureau Statistics (ABS) 2011. 'Characteristics of recent migrants'. In *Perspectives on Migrants, 2011*. ABS Catalogue No. 3416.0. Canberra: ABS.

Baldassar, L. 2001. *Visits Home: Migration Experiences between Italy and Australia*. Carlton South, Victoria: Melbourne University Press.

Baldassar, L. 2007. 'Transnational families and aged care: the mobility of care and the migrancy of ageing'. *Journal of Ethnic and Migration Studies*, 33(1): 275–297.

Baldassar, L., Baldock, C.V. & Wilding, R. 2006. *Families Caring Across Borders: Migration, Ageing and Transnational Caregiving*. Basingstoke: Palgrave MacMillan.

Beaverstock, J.V. 2005. 'Transnational elites in the city: British highly-skilled inter-company transferees in New York city's financial district'. *Journal of Ethnic and Migration Studies*, 31(2): 245–268.

Benson, M.2011*The British in Rural France: Lifestyle Migration and the Ongoing Quest for a Better Way of Life*. Manchester: Manchester University Press.

Benson, M. & O'Reilly, K. 2009a. (eds) *Lifestyle Migration: Expectations, Aspirations and Experiences*. Farnham, Burlington: Ashgate.

Benson, M. & O'Reilly, K. 2009b. 'Migration and the search for a better way of life: a critical exploration of lifestyle migration'. *The Sociological Review*, 57(4): 608–625.

Benson, M. & Osbaldiston, N. (eds) 2014. 'New Horizons in lifestyle migration research: theorising movement, settlement and the search for a better way of life'. In *Understanding Lifestyle Migration: Theoretical Approaches to Migration and the Quest for a Better Way of Life*. London: Palgrave Macmillan.

Blunt, A. & Dowling, R. 2006. *Home*. London: Routledge.

Bourdieu, P. 1983. 'Site effects'. In *The Weight of the World: Social Suffering in Contemporary Society*, edited by P. Bourdieu, 123–130. Stanford, California: Stanford University Press.

Clarke, N. 2004. 'Free independent travellers? British working holiday makers in Australia'. *Transactions of the Institute of British Geographers*, 29(4): 499–509.

Clarke, N. 2005. 'Detailing transnational lives of the middle: British working holiday makers in Australia'. *Journal of Ethnic and Migration Studies*, 31(2): 307–322.

Clifford, J. 1986. 'Introduction: partial truths'. In *Writing Culture: The Poetics and Politics of Ethnography* edited by J. Clifford and G.E. Marcus, 1–26. Berkley: University of California Press.

Conradson, D. & Latham, A. 2005a. 'Friendship, networks and transnationality in a world city: Antipodean transmigrants in London'. *Journal of Ethnic and Migration Studies*, 31(2): 287–305.

Conradson, D. & Latham, A. 2005b. 'Transnational urbanism: attending to everyday practices and mobilities'. *Journal of Ethnic and Migration Studies*, 31(2): 227–233.

Douglas, M. & Isherwood, B. 1978. *The World of Goods: Towards an Anthropology of Consumption*. Harmondsworth: Penguin.

Dowling, R. & Power, E. 2012. 'Sizing home, doing family in Sydney, Australia'. *Housing Studies*, 27(5): 605–619.

Dupuis, A. & Thorns, D.C. 1996. 'Meanings of home for older home owners'. *Housing Studies*, 11(4): 485–501.

Easthorpe, H. 2014. 'Making a rental property home'. *Housing Studies*, 29(5): 579–596.

Favell, A., Feldblum, M. & Smith, M.P. 2007. 'The human face of global mobility: a research agenda'. *Society*, 44(2): 15–25.

Forrest, J., Johnston, R. and Poulsen, M. 2014 'Ethnic capital and assimilation to the Great Australian (homeownership) dream: the early housing experience of Australia's skilled immigrants'. *Australian Geographer*, 45(2): 109–129.

Ganga, D. and Scott, S. 2006. 'Cultural "insiders" and the issue of positionality in qualitative migration research: moving "across" and moving "along" researcher-participant divides'. *Forum Qualitative Sozialforschung / Forum: Qualitative Social Research*, 7(3): Art 7.

Hammerton, A.J. and Colborne, C. 2001. '"Ten-pound poms revisited: battlers" tales and British migration to Australia'. *Journal of Australian Studies*, 25(68): 86–96.

Hammerton, A.J. and Thomson, A. 2005. *Ten Pound Poms: Australia's Invisible Migrants*. Manchester: Manchester University Press.

Hollinsworth, D. 1998. *Race and Racism in Australia*, 2nd edn. Social Science Press: Katoomba NSW.

Hugo, G. & Harris, K. 2011. 'Population distribution effects of migration in Australia'. Canberra: Department of Immigration and Citizenship.

Iredale, R. 2005. 'Gender, immigration policies and accreditation: valuing the skills of professional women migrants'. *Geoforum*, 36(2): 155–166.

Jacobs, J.M. 2006. 'Too many houses for a home: narrating the house in the Chinese diaspora'. In online papers archived by the Institute of Geography, School of Geosciences, University of Edinburgh.

James, C. 2009. 'Australian Homes are the biggest in the world'. *Economic Insights*, Commsec. 30 November.

King, R. 2002. 'Towards a new map of European migration'. *International Journal of Population Geography*, 8(2): 89–106.

Kofman, E. 2012. 'Rethinking care through social reproduction: articulating circuits of migration'. *Social Politics*, 19(1): 142–162.

Lunt, N. 2008. 'Boats, planes and trains: British migration, mobility and transnational experience'. *Migration Letters*, 5(2): 151–165.

Mar, P. 2005. 'Unsettling potentialities: topographies of hope in transnational migration'. *Journal of Intercultural Studies*, 26(2): 361–378.

Martin, G. 2015. 'Stop the boats! Moral panic in Australia over asylum seekers'. *Continuum*, 29(3): 304–322.

Miller, D. 1998. *A Theory of Shopping*. Cambridge: Polity Press.

Miller, D. 2010. *Stuff*. Cambridge: Polity Press.

Office of National Statistics (ONS) 2013. *A Century of Home Ownership and Renting in England and Wales*. [Online] Available at: http://webarchive.nationalarchives.gov.uk/20160107120359/http://www.ons.gov.uk/ons/rel/census/2011-census-analysis/a-century-of-home-ownership-and-renting-in-england-and-wales/short-story-on-housing.html.

O'Reilly, K. & Benson, M. 2009. 'Lifestyle migration: escaping to the good life'. In *Lifestyle Migration: Expectations, Aspirations and Experiences*, edited by K. O'Reilly & M. Benson, 1–13. Farnham, Burlington: Ashgate.

Phillips, J. & Spinks, H. 2012, *Skilled Migration: Temporary and Permanent Flows to Australia*, Background note, Parliamentary Library, Canberra, 6 December 2012.

Richards, L. 1990. *Nobody's Home: Dreams and Realities in a New Suburb.* Oxford: Oxford University Press.

Salt, B. 2012. 'The suburb with 43pc Poms: it's our own Little Britain'. In *The Australian*, 7 August.

Scott, S. 2006. 'The social morphology of skilled migration: the case of the British middle class in Paris'. *Journal of Ethnic and Migration Studies*, 32(7): 1105–1129.

Sriskandarajah, D. & Drew, C. 2006. *Brits Abroad: Mapping the Scale and Nature of British Emigration.* London: Institute for Public Policy Research.

Voloder, L. 2008. 'Autoethnographic challenges: confronting self, field and home'. *The Australian Journal of Anthropology*, 19(1): 27–40.

Walsh, K. 2011. 'Migrant masculinities and domestic space: British home-making practices in Dubai'. *Transactions of the Institute of British Geographers*, 36(4): 516–529.

Wiles, J. 2008. 'Sense of home in a transnational social space: New Zealanders in London'. *Global Networks*, 8(1): 116–137.

4 British migrant orientations in Singapore

Negotiating expatriate identities

Sophie Cranston

Do you consider yourself an expatriate?
  'I never really considered too much what it means, but the way that people refer to us is as expats probably, you know, I consider myself in a group of other people that are generally referred to as expats.'
                                        (Theo,[1] 6 months in Singapore, April 2012)

For the roughly 60,000 British people living in Singapore in 2013 (Finch et al., 2010), there were a wide range of reasons as to why they had moved there. Some are sent on secondments by the organisation that they work for, others find a job and head over, others go to find a job, some are in search of a better lifestyle, others are looking for opportunities for their children, some are accompanying or following their partners. Many don't move directly from the United Kingdom, some have spent most of their lives abroad, while others intend to return to the UK, some have settled in Singapore, and many think about moving somewhere else afterwards—Australia, Hong Kong, France. Highlighting these different journeys to, through, and from, Singapore, shows that migration is not a singular event. The time spent by British migrants in Singapore is one part of a much longer migrant journey through space and time (Schapendonk & Steel, 2014). It is therefore not a step where, upon stepping off the plane at the airport, the British migrant becomes 'expatriate.' The negotiation, practice and articulation of their migrant identities is an ongoing and changing process.

The expatriate as a migrant identity is one that is often axiomatically applied to British migrants. While the term 'expatriate' acts as a common nomenclature denoting a skilled migrant who lives abroad for a temporary period of time, it is also imbued with a 'western and national baggage' and thus is associated often exclusively with privileged white migrants (Cranston, 2017, Fechter, 2007, Knowles & Harper, 2009, Leonard, 2010). I do not by any means suggest that the exclusivity of the ways in which the term expatriate is employed is unproblematic. However, in this chapter I use discussions around this term by British migrants to highlight the way in which the 'expatriate' becomes associated with certain practices. In doing so, I draw upon research in the field of mobilities which argues that we need to move beyond place-based accounts to think about how mobilities—including migration—are produced relationally to one another (Cranston, 2017; Hui, 2015; Glick-Schiller & Salazar, 2013). I argue that this research should necessarily extend into exploring how British migrant identities are produced in relation to other British migrant identities. This involves extending an understanding of the 'social fissures' (Scott, 2004) that make up migrant communities, to think more seriously about the complexity of mobile migrant lives.

A way through which we can understand the complexity of British migrant lives abroad is by looking at practices. Practices refer to the 'panopoly of mundane efforts' (Conradson & Latham, 2005, 228) through which people negotiate their (migrant) lives. These include things like eating, drinking and socialising which are social and material practices that constitute the experience of migration (Ho & Hatfield, 2010). Through this chapter, I focus on how certain practices become associated with 'expatriate' ways of being on British migrant journeys through Singapore. Situated within calls to 'ground' research on transnationalism, research on practices in the context of expatriate migration highlight how meaning is produced through the everyday (Dunn, 2010; Ho & Hatfield, 2010; Walsh, 2006). This work looks at practices associated with encounter, how expatriate identities are shaped in relation to what they are not in the place of stay abroad. For example, Coles and Walsh (2010) look at dress, food and excursions in Dubai, exploring how these 'reassert a seemingly clear distinction between the cultural practices of Self and Other' (p. 1323). Identity here is seen as a construct that is produced through the emotional response to being in a different culture, with practices being negotiated through this response. As an example, we can see this identity-making through research that has been carried out on one practice—club membership. Beaverstock (2011) highlights how British migrants in Singapore utilise British club spaces as a 'sanctuary' from other parts of city; Knowles and Harper (2009) highlight how 'expat' clubs in Hong Kong are 'places to socialise, eat, and drink with like-minded people' (p. 199). Therefore, through practices like club membership we can see how identities are both expressed and produced, how belonging is embodied and material, for example, through retreating from other spaces of the city. Hence, we can explore an understanding of lifestyle in migration, showing how migrant lives play out through consumption (Benson & O'Reilly, 2016).

In this chapter I extend this previous research on how practices have meaning in producing migrant identities. However, rather than looking at how practices work to make distinctions between British migrants and the 'Other' as the culture of the abroad, or as an assertion of British lives abroad, the chapter looks at how British migrants use practices to make distinctions between an 'expatriate' identity and different migrant lives. I argue that practices become a way in which white British migrants distinguish themselves from other white British migrants 'in the absence of visible and supposedly immutable signs of difference such as 'race' or 'skin colour' (Yeoh &

Willis, 2005, p. 277). That is, as white British migrants are assumed to all be the same through their skin colour by outsiders, practices such as whether or not they brunch become ways in which distinctions can be made within the community. Through the chapter, expatriate is used mostly in a non-nationalistic sense. The British migrants I interviewed didn't articulate different practices of British or American expatriates, but used the term expat to distinguish between groups of Western migrants more generally. Although discussions of practices associated with differences between the Singapore and the UK cannot be generalised to other groups of migrants (for example, climate and weather is distinctively British), the practices themselves (holiday lifestyle, brunch, travelling, charity work) are more widely associated with Western migrants under the umbrella term of 'expatriate.'

Certain elements of Singapore inform the practices of the British migrant in this location—it is an Asian country, a former British colony, a city-state marked by transience and immigration laws that encourage 'foreign talent' which Britons by virtue of their nationality feel that they fall under. Singapore's migration regime has been explored in depth (see, for example, Yeoh, 2012), with British migrants in Singapore occupying a privileged position in terms of their immigration status as desirable to the long-term development of Singapore's knowledge economy. However, through the chapter, I argue that this acts as a background as opposed to determining feature to the ways in which British migrant identities are understood in relation to other British migrants, as most British migrants occupy similar visa statuses.[2]

The chapter draws upon 36 interviews carried out with 39 British migrants and participant observation carried out between February and April 2012.[3] Out of the 39 respondents, 36 are white British, but the race of the respondents is not highlighted in the chapter to preserve anonymity of the non-white participants. The working-aged migrants who make up the sample had been abroad between 3 weeks and 30 years, working in a variety of sectors including finance, education, construction, shipping and law. The variation of time abroad was important to note in the ways in which these British migrants understood their lives abroad—the overall goal of the wider research project was to explore the multiplicity of the ways in which the term 'expatriate' is produced and consumed. This chapter focuses on the ways in which these British migrants expressed their status as expatriate or not expatriate, to see how British migrants understood the figure of the expatriate which they are often assumed to inhabit. In some ways, as a British migrant in Singapore for a short period of time, I occupied a similar positionality to that of my respondents, however, I didn't see myself as an expatriate. For the British migrants I interviewed, there was an ambivalence towards, active identification with and rejection of the term to describe their lives in Singapore. For some, such as Theo above, it was a term that they hadn't really thought about, it was just a name that they saw as being applicable to people like them. Within the answers there were three overlapping narratives that discussed the expatriate in terms of race, lifestyle and citizenship. Through this chapter, I examine practices that British migrants articulated around lifestyle—drinking, eating, going out; and citizenship—visas, travelling and passports.[4] I focus specifically on how British migrants articulated their own identity as 'expatriate' or something else, as opposed to looking at how this identity becomes ascribed to certain bodies. This therefore looks at how people perform their identity through practices. However, I look at how the expatriate acts as an *identity-orientation* through which British migrants understand their lives in Singapore. As an identity-orientation, the expatriate becomes something that individuals fit with or against. Rather than a self-evident identity, the expatriate is an identity that migrants orient themselves towards through certain practices. That is, the individual can reject, turn away, or orient themselves away from the expatriate and orient themselves towards a different identity.

The chapter first presents a discussion of the ways in which British migrants actively produced themselves as expatriate or not expatriate in relation to three lifestyle practices—brunch, lunch and lying by the pool. It argues that these are three practices that portray 'expatriate' life as being like a hedonistic holiday that some British migrants actively seek to distance themselves from. Second, the chapter looks at different articulations of British migrant belonging to Singapore, again highlighting how this is discussed through practices of citizenship such as passports and travelling, as well as other practices of belonging such as charity work.

Lifestyles

SC: What is your image of an expatriate?
GEORGE: It is a vision of someone sitting by the pool enjoying their life in a foreign country.(10 years in Singapore, March 2012)

Colonial imaginaries about expatriates are evoked through the practice of sitting by the pool (c.f. Fechter, 2007). The image of the expatriate sipping their gin and tonic by the pool at sunset was one that my British migrant respondents, like George articulated and James played with: 'as late as sunset [laughs]' (7 years in Singapore, 12 years abroad, March 2012). Guides to expatriate life suggest that it is normal for British migrants to live in condominiums with pools, gyms, tennis and squash courts (Cranston, 2016a). However, British migrants explained this as being part of the Singaporean lifestyle—being by the pool was less of a marker of difference from Singaporean culture, than from life at home. For example, as James continued, expats: 'sit around the pool and their gins at sunset and that's not because they are expats it is because they are here in a climate and society where you can do that kind of thing. Yeah?' (March 2012).

This example of everyday life from British migrants in Singapore highlights what some perceive as being an artificial lifestyle. For some British migrants in Singapore, there is a sense that life for the expatriate is not quite real—it is not real life because it is like a holiday. For example, this was present in Liam's discussion of whether he was planning on returning to the UK: 'So maybe we are lucky because we have always got that element of it feeling a little bit like a holiday even though you know in reality its normal life' (Liam, 9 years in Singapore, March 2012). Part of this holiday vibe was attributed to the 'condo lifestyle':

> condo lifestyle was great in that it felt you were on holiday because you have got your own pool and you have got your own tennis court and there's a guard at the gate and you know these kinds of things, squash courts and gyms and all that, but it is really, is kind of like being on holiday.
>
> (Thomas, 3 years in Singapore, 6 years abroad, March 2012)

As John Urry and Jonas Larsen (2011) suggest, holidaymaking is seen as a departure, a 'limited breaking with established routines and practices of everyday life and allowing one's senses to engage with a set of stimuli that contrast with the everyday and mundane' (p. 3). However, as practices like lying by the pool become seen as a new routine, a 'quiet day activity' (Jake, 1 year in Singapore, March 2012), following Urry and Larsen (2011), we can see that it no longer marks the deviance from everyday life that we associate with the tourist gaze. Instead for British migrants, lying by the pool, rather than a departure from the mundane associated with tourism, comes to indicate a form of deviance from life in the UK. Lying by the pool becomes a way in which British migrants articulate what it means to be a migrant in Singapore, a lifestyle that becomes labelled as expatriate.

Being or not being an expatriate for Western migrants however was most actively articulated around the practice of brunch. This was something that Hannah, who considered herself to be an expatriate, naturalised as being almost routine: 'a certain lifestyle, so you go to the brunches and there are certain places you go to, umm, yeah it's a way of life, it's just this kind of bubble, yeah' (Hannah, 18 months in Singapore, April 2012). In the Singaporean context, brunch involves going to a high-end Singaporean bar or hotel where you pay a flat fee, often up to S$150 (roughly £75 in 2012) and you can have as much food and drink as you want, what is described as 'free-flow.' This practice is very different from brunch, eggs benedict and a cup of coffee, in a UK setting. Practising brunch then is a form of orientation to a lifestyle that is labelled as expatriate within Singapore. Although lying by the pool with a gin and tonic is also a practice associated with the better off, brunch comes to represent the 'lavish lifestyles, hedonistic pleasures' (Leonard, 2010, p. 1) that are commonly associated with expatriates. This means that this lifestyle is something that British migrants actively rejected through their articulations of whether they considered themselves as expatriates or not: 'I'm not an expat. I don't brunch. I'm not paying S$100 per person so I can have a free flow of alcohol when I have two kids. I don't want to have a hangover during daylight' (Field notes, March 2012). Through this man's discussion we can see that some practices are exclusionary for some groups of British migrants. For example, there are some British migrants who are unable to afford the high price tag, as traditional expatriate packages which offered additional benefits are increasingly no longer offered to migrants in 'attractive' destinations like Singapore (Cranston, 2016b). Others, such as those with children, like the example of this man, do not see this as part of a family friendly lifestyle in Singapore.

The criticism of certain lifestyle practices by British migrants who don't consider themselves as expatriates, often acted as a critique of people who did practice this lifestyle. We can see this most clearly in the example of 'ladies who lunch':

> 'We are the typical expat women' they say. We have too much fun. Yes, our fun generally starts at 12 over there with giant margaritas and lunch [pointing to the Oyster Bar across the Bay]. It is better to be a housewife in Singapore than in the UK, they continue, because in the UK their husbands are away from 8 to 7 and while they may also be away here, they have more people to play with.
>
> (Field notes, March 2012)

This self-parody of ladies who lunch by two women that I met at an international women's coffee morning illustrates the comparisons that are made between being housewives in the UK and housewives in Singapore—with women's lives in Singapore being marked as play and Singapore the playground. This also illustrates the ways in which 'expatriate' practices are gendered, with an assumption of British migrant men being the lead migrant, followed by their 'trailing spouse' (Fechter, 2010; Leonard, 2010; Walsh, 2008). Despite the fact that this does not reflect contemporary migration trends, the dominant representation of 'expatriate' women is of 'someone with plenty of time and money on her hands to devote to leisure pursuits' (Leonard, 2010, p. 105). For the two British women I was speaking to, this focus on fun, going out drinking at lunchtime, was a typical practice of the typical expatriate woman. It is also one that we can associate with ideas of privilege, it is a lifestyle that excludes others on the basis of wealth—both in terms of the ability to afford being in Singapore on a single income and the price tag of places like the Oyster Bar. The 'ladies who lunch' as a description, is an often derogatory way to describe female British migrants who are perceived to live a life of leisure. Therefore, the 'ladies who lunch' lifestyle—with practices of lunch, tennis and clubs, generally acted as a critique to which other (often male) British migrants compared themselves, or their family, too. This referenced what Meike Fechter (2010) describes as the 'myth of the lazy woman,' highlighting the supposed idleness and hedonism of female Euro-American migrants, particularly in terms of thinking about consumption (see also Walsh, 2008). Elliot also suggested this, in his comparison between a lady who lunches and what he portrays as 'worthwhile' activities—work or charity work:

> They go from their home to Tanglin Trust School, which is basically the British school, drop off their kids, if they don't maybe the domestic worker does it and then they go to the club whether it is, in fact in this instance it is the Dutch club [pointing to the club next door to the interview location], they go play tennis three times a week there, they have lunch with their friends and then they go home and they do this very colonial, live this very colonial lifestyle.
>
> (Elliot, 8 years in Singapore, 21 years abroad, March 2012)

Here, Elliot was trying to explain why he feels that a friend of his, who he described as a 'certain type' of expatriate, claimed not to have heard of parts of Singapore outside the centre because they live a lifestyle that is detached from it. This lifestyle practice is something that he also describes as being 'very colonial.' Critiquing ladies who lunch is a way he articulates that he is not an expat and distinguishes himself as a more modern type of migrant who has a different engagement with Singapore.

The way in which British migrants articulate a sense of expatriate life as being like a holiday, artificial, worked as a way of distinguishing it from being different to everyday life in the UK. In this way, for British migrants, being an expatriate became seen as practising a different lifestyle from that in the UK. Expatriate was a label ascribed to a set of practices that worked to make distinctions. We can see how this relates to wider understandings of migration and tourism, as Karen O'Reilly suggests in the context of British migration on the Costa Del Sol that ideas about the migrant 'are linked to notions and judgements about the modern tourist ... the image of this sort of traveller is one who is ... hell bent on hedonism, spending, freedom and indulgence' (2000, p. 18). In their discussions of what it means to be an expatriate, similar ideas about hedonism are expressed by British migrants in the ways through which they try and make sense of what they perceive to be an artificial and unreal existence: 'It's not real life. We are all aware that's it not real life. You are in a bubble' (Sam, 18 months in Singapore, March 2012). In this way, we can see that lifestyle practices became a way in which British migrants could distinguish themselves from other British migrants. For those British migrants who rejected it, instead of the expatriate they would orient themselves towards a different lifestyle.

Rejecting the label expatriate therefore is a way through which British migrants claimed different types of belonging to Singapore. The rejection of the term expatriate here can be read as a claim for the 'ordinary' as opposed to the hedonistic. This group of migrants were tired of what Emily described as the 'Groundhog Day' of going out and drinking every night (1 year in Singapore, 3 years abroad, March 2012). In rejecting the term expatriate, they drew upon Bourdieuan notions of taste in order to make classifications as to what constitutes the expatriate whereby they 'distinguish themselves by the distinctions that they make ... the distinguished and the vulgar, in which their position in the objective classifications is expressed or betrayed' (Bourdieu, 1984, p. 5). For them, the problem wasn't necessarily that the expatriate lifestyle was high end, but that it was decadent: 'But I know people who do that every week, that is their Sunday, that is what they do. And they go around the island trying different brunches, that's ridiculous, it's just indulgence' (Liam, 9 years in Singapore, March 2012). The comparison between their practices to practices that they considered as being expatriate, worked to produce notions of British migrants who had been in Singapore for a shorter amount of time as being like mass tourists, a different type of people who were only interested in themselves and their leisure. This is something that Thomas clearly highlights by referencing living in a condo as resembling going on a package holiday: 'The city square for example, in, umm, Little India that is just you know like going on a package holiday with Thomson which is like there's a massive pool, there are loads of people there' (Thomas, 3 years in Singapore, 6 years abroad, March 2012). Or one long-term British migrant lamented the recent increase in British migrants as turning Singapore into the 'new Costa del Sol' (Field notes, March 2012). The British migrants who brunch therefore are criticised as both lacking adventure and only being interested in Singapore as a playground for a temporary period of time:

> Otherwise you are not going to get into that system and not have that sense of ownership to be here for a bit longer. And on Sunday at 10.30, you can see it a mile away, the people who just do brunch. That's their morning every weekend. What do they get out of that really? What the hell else do they do in Singapore? These people who do brunches in high end bars because they don't go elsewhere. These are the people on two year contracts then they are out.
>
> (Michael, 7 years in Singapore, 21 years abroad, March 2012)

However, the ways in which British migrants in Singapore articulated different forms of belonging works to highlight a multiplicity of identities. Although British migrants stated that because they didn't engage in certain practices, for example brunch, meant that they didn't consider themselves expatriates, they also found it difficult to articulate a different category of migrant belonging:

SC: So if you don't want to be called an expat, what would you like to be called?

LEWIS: That's a good question actually.

> (5 years in Singapore, 22 years abroad, April 2012)

No respondents described themselves as immigrants or migrants, which can be seen a reflection of wider debates about race and migration in both a UK and Singaporean context (see also Cranston, 2017). No respondents described themselves as Singaporean either, even the one respondent with Singaporean citizenship. Therefore, in claiming and practising different types of belonging to Singapore, some respondents utilised a terminology which highlighted their nationality as opposed to migrancy, describing themselves as a 'foreigner' or a form of British belonging, such as 'Scottish.' However, as highlighted above, because white British migrants are seen to fit the category of expatriate, despite their protests and desire to call themselves something different, this is often what they are axiomatically assumed to be: 'if you are white in Singapore then you are, the first thing that someone is going to say or think when they meet you is, expat' (Emily, 1 year in Singapore, 3 years abroad, March 2012). This is a way through which 'expatriate' acts as a raced category (Cranston, 2017; Fechter, 2007; Knowles & Harper, 2009; Leonard, 2010), one that again becomes associated with certain practices. This is something that Colin highlighted through his discussion of changes he saw in the British migrant community in Singapore: 'possible to live here and not to experience Singapore, with the clubs, Holland Village, met neighbours in condos, swimming pools—they are all Caucasian' (Colin, 40 years in Singapore, March 2012). His use of 'Caucasian' here illustrates the way in which the practices of detachment and hedonism described above are associated with the white British migrant.

For those British migrants who want to orient themselves away from an expatriate identity, one way this became expressed was by articulating a different migrant identity—the Ang Moh. Ang Moh, translated as red face or red hair, originally seen as a derogatory Hokkein term for a white person, like the bule in Bahasa, or the gweilo in Cantonese, is a way of describing a migrant belonging to Singapore. For example, George when asked how he would describe himself suggested that: 'The locals amongst us call ourselves ang-moh porians. Ang Moh is what the Singaporeans call us, which is technically racist. However, we call ourselves Ang Moh porians. There's no offence meant by it' (10 years in Singapore, March 2012). Through this, he is highlighting a different kind of belonging to Singapore than the people who engage in the pool side life. They are still migrants but highlight a distinct Singaporean migrant identity. This is something that Liam also highlighted, again using Ang Moh as a way of distancing himself from expatriates who brunch: 'Now its 9 years [length of time in Singapore] everybody is like oh sure the person you should talk is [name], he is always here, he's going nowhere. Umm but that hasn't turned around for the locals, they still look at me like the crazy Ang Moh' (Liam, March 2012). Therefore, using the term Ang Moh instead of expat can be seen to be more playful, an acknowledgement that you are marked as different but an attempt to distance the self from the representations, and practices, associated with the expatriate. In particular, as Thomas highlights, it suggests an engagement with practices that are seen as more local in the Singaporean context, for example, living in different types of residences without the swimming pool or hanging out with friends in the Hawker centre [open air food court]: 'hanging out in Hawker centres and I do that, go to Hawker centres and meet people, you go there for cheap beers' (Thomas, April 2012). For these British migrants, these claims are not about being excluded from practices in terms of cost, or family life, but more about practising a more local type of belonging, using a migrant identity that is specific to Singapore.

Therefore, as this section has shown, one of the ways in which British migrants reflect on not being expatriates is through practices associated with lifestyle— brunch, lounging by the pool, going for lunch. They utilise these practices as a method of making distinctions between being an 'expatriate'—an identity that they critique as hedonist, linked to mass tourism and being distant from the local population—and themselves. As the quotes from my respondents illustrate, these critiques are offered by those migrants who have been abroad for longer periods of time—from 3 to 40 years. By not spending every weekend going to brunch or lying by the pool, meant that these respondents didn't consider themselves to be expatriates. However, there was not a clear sense of other identity-orientations that they utilised to make sense of their lives abroad. The exception to this was the drawing upon of practices associated with the Ang Moh which was a more locally engaged and playful way through which they articulated what it means to be a British migrant in Singapore. This identity-orientation was again expressed through different social practices, such as going to the Hawker centre for a beer, instead of the high-end bar.

Citizenship

As the discussions above highlight, it is the British migrants who have been abroad for longer periods of time that are more likely to orient themselves away from the expatriate. In many respects, this fits with commonplace understandings of the expatriate, which can be defined with a notion of return. This idea of return manifests itself within practices of citizenship— including visas, passports as well as the retention of property in the UK. As Burrell (2008) suggests passports and visas illustrate the materiality of migrant journeys, a point that can be applied more broadly to the ways in which we understand the materiality of migrant practices. This adds another layer of complexity to the ways in which we understand the expatriate as an identity-orientation, as some of my respondents in discussing why other British migrants rejected the term expatriate, in turn, rejected this rejection:

> but they are British and they say that they are not an expat because this is their home, but they still have British passports, but they are still British citizens, they are not Singaporeans but they have the right to live here for a long period of time. That still classes them as expat in my book.
>
> (Luke, 2 years in Singapore, 15 years abroad, March 2012)

For some of my respondents to define what they felt an expatriate to be, they provided a scale that looked at immigration status in Singapore:

> So on a one to ten scale; I'd say I'm a four, with six being a non-expat. So I'm half way between. The way I see it, there is a middle ground between being an expat and a citizen. The middle is PR [Permanent Resident]. That's where I see myself. An expat is on an EP [Employment Pass] and here for a few years, for citizens this is home.
>
> (Ash, 3 years in Singapore, March 2012)

Ash suggested that this scale was based on three immigration categories—the employment pass, a two-year work permit tied to an employer; permanent residency, a five-year residency visa and citizenship. For Ash, being on an employment pass, a visa that lasts two years and is tied to your job meant that you were an expatriate, because of its temporality and the implied intention to go home afterwards. This meant that one of the defining features of the 'expatriate' experience was the knowledge that circumstances might change, and thus the need to enjoy yourself while it lasted: 'So yeah so what I've found with being an expat is try and enjoy yourself, ultimately, but don't get too comfortable though ... because one day you may find that actually your job is passed to a local hire' (Sam, 18 months in Singapore, March 2012). In some ways, we can see how this is linked to understandings of hedonistic tourism—the idea of making the most of time while you are in Singapore. Ash was articulating ideas of the expatriate as belonging as a transient state, which resulted in either leaving Singapore, or applying for Permanent Residency or Singaporean citizenship.

However, there is little reflection here on how citizenship is entangled with understandings of the lifestyle that material practices of citizenship enable, illustrating that they are 'the people who have the resources, variously of time, money, credentials to undertake these journeys' (Amit, 2007, p. 2). In describing their lifestyle in Singapore, respondents highlighted travel elsewhere as an important part of it: 'Whereas here I will go away I will go this weekend and jump on a plane and go to Hong Kong or jump on a plane and go to Thailand, or Bali' (Jade, 1 year in Singapore, March 2012). Or Emma who was discussing her feelings about being on a dependent pass associated with her husband's job: 'it is not preventing me from really doing anything, I can move around and travel, I went to Thailand recently for a few days with friends a couple of weeks ago' (6 weeks in Singapore, March 2012). The ability to travel easily around the Asia-Pacific region is a practice associated with their British passport, which often enabled visa and hassle-free travel for leisure. Indeed, for some, their decisions about whether to take up citizenship were linked to the ways in which they understood the passport as a document. Singaporean nationality for Nathan was seen in terms of being a travel document, one which would confer the opportunity to travel around the region on business without requiring a visa: 'do I want a Singaporean passport? No. I have quite a few friends who have got their Singaporean nationality ... [the friend] asked me the same thing, am I going to get mine, but why would I?' (Nathan, 7 years in Singapore, 11 years abroad, April 2012).

The decision to apply for Permanent Residency status was however based upon the perceived benefits that this would entail, a solution against the feelings of transience. For example, Ash was considering applying for PR status because he recognised that if he lost his job he would have 30 days to leave (March 2012). It became seen as a specific form of capital that enabled a sense of security in Singapore, based upon the perceived 'benefits' that this might offer. While most British migrants didn't question, or want to change, their British citizenship, the expatriate became a marker for how they viewed their status as a temporary migrant with this having a materiality in expression through visas. For example, George felt comfortable being a 'foreigner' as 'I've got no real desire to become a PR to be honest, I don't see that many benefits' (March 2012). For some, the mobility that the British passport enabled was sufficient, one that materially marked their status as not belonging in Singapore. Therefore we can suggest that rather than act as a form of transnational, flexible citizenship which sees the migrant 'acquiring a range of symbolic capitals that will facilitate their positioning, economic negotiation and cultural acceptance in different geographical sites' (Ong, 1999, pp. 17–18), Permanent Residency was a specific form of capital that enabled a sense of security in Singapore. This sense of security enabled British migrants to make more long-term plans to stay in Singapore. This is again another way in which attachments to Singapore highlight an alternative way of understanding British migrant identities.

For others, the decision to get PR was embedded within the additional responsibilities to Singapore which it entails. We can see this explicitly through discussions around national service. The permanent residency in Singapore of male British migrant young people through their parents means that they become obligated to undertake Singapore's national military service at 19. Joanna who had just obtained PR status at the time of interview reflected upon this:

> It does mean we have just signed two years of [son]'s life away for national service, but again, I think we are kind of not your typical expats with that, I don't have a huge objection to him doing national service ... I have got a friend of a friend who has been here for 18 years and her husband just took her son back to the States so he doesn't have to do national service and think ... morally, I can't do that. I can't say then, but oh you are different, you don't have to do that.
>
> (Joanna, 2 years in Singapore, April 2012)

Joanna then felt that her willingness for her son to carry out national service marked her out as being a different type of expatriate. She described this as not being 'typical,' because she wasn't willing to shirk the responsibilities that came with being a permanent resident. She contrasted this with Western migrants simply using the PR system as a sense of security. For the British migrants who didn't associate with the lifestyles associated with temporality and hedonism then there were different practices of belonging. The rejection of the term expatriate was often a call of being more culturally engaged and aware of their surroundings:

> We are here and giving back to Singapore, I do so many different local concerns, I do riding for the disabled here, I do reading for the blind ... we are not a member of the clubs here that hang out with the other expats, we know so many Singaporean people and we count Singapore as our home and we are proud to do so. And I think Singapore has a lot of pluses. I wouldn't count of myself as being Singaporean, but I wouldn't count myself as an expat.
>
> (Abigail, 7 years in Singapore, 20 years abroad, April 2012)

Again, discussions of being *not* an expatriate were used by British migrants to highlight different types of belonging to Singapore. As previous research has highlighted, 'home' and 'belonging' indicate practices that play on emotional registers of identity (Walsh, 2006). For Abigail, the way in which she expressed that she wasn't an expatriate was through highlighting her civic engagement in Singapore through charity work, an engagement which enabled her to get to know the country: 'And I think we are richer people for knowing Singaporeans, than not, I think that is why we are not expats. And proud not to be expats' (April 2012). In many ways, this is an interesting interpretation of the role that charity work plays in the constitution of identity. For example, in returning to the self-parody of 'ladies who lunch,' the feminisation of this practice is due to portrayal of the lead migrant being a male professional, accompanied by his 'trailing spouse' (Berry & Bell, 2011). Here, the (assumed female) spouse is often seen to engage with a third sphere of community work as a substitute for paid employment (Yeoh, et al., 2000). However, here Abigail (who has a full-time job) upsets this notion by articulating that she is not an expatriate because she is engaged in community work and expats are more interested in play—being down the club for example. This once again highlights that while being an expatriate is associated with being in a bubble, artificial or on holiday, other types of belonging were articulated through practices that highlight a 'sense of attachment' to Singapore as a place. These different types of belonging are ones which we can see as being produced over longer periods of time (Walsh, 2011, p. 54). This understanding of engagement was not just highlighted through charity work. For example, one of the ways in which Andrew articulated that he was an expatriate was through the lack of political engagement that this enabled him to pursue. One of the reasons that Andrew had spent his life 'expat-ing' was 'I enjoy the, the same sort of lack of responsibility in a way, I find it easier in a way' (1 year in

Singapore, 23 years abroad, March 2012). Even though Andrew criticised the lack of freedom of speech in Singapore he shrugged this off by saying:

> But it's not my country, get on with it. If that's where you want your country to be, fine I will go along with it until I can't stand it any longer and I will say thank you very much and good night.
>
> (March 2012)

For him, part of being an expatriate in Singapore was about not practising politics, a view echoed by research carried out on Western 'expatriates' in the Netherlands (van Bochove & Engbersen, 2015). Enjoying the transience that his life as an expatriate offered, Andrew's option as an expatriate was to either cope with what he felt to be problematic policies or to leave. Therefore, we can see then that for some British migrants the rejection of the term expatriate was a way through which they articulated life in Singapore as being real life, about being engaged to that place.

This section has illustrated other ways in which British migrants in Singapore orient themselves towards or against the expatriate by looking at practices associated with citizenship. These practices are both material, in particular, thinking about passports and visas, but also social, for example, travel, charity work and the engagement in politics. In some ways, these practices, 'fit' with the definitions of the expatriate highlighted in the lifestyle section. For example, being on an employment pass enables you be in Singapore for two years, meaning that British migrants want to make the most of their life while abroad. Being a permanent resident, even if this visa only lasts for five years, begins to highlight a different kind of belonging to Singapore, as suggested within its name: permanent. But, in other respects, thinking about practices associated with citizenship also helps highlights the contradictions of the way in which the expatriate as an identity-orientation is understood by British migrants. For example, not taking up Singaporean citizenship, for some, suggests that you always see your presence in Singapore as being temporary even if you practice a lifestyle that seeks to highlight your engagement with the place.

Conclusions

This chapter has looked at the ways in which social and material practices—brunch, lying by the pool, where you eat lunch, charity work, travel, visas, passports—act as ways in which British migrants in Singapore orient themselves towards and against expatriate identities. In this way, these practices mark ways in which the British migrants who inform my research make distinctions within Singapore. Importantly, these distinctions are not only ways in which they work to articulate a sense of their migrant lives, but how they make sense of their lives in Singapore as being different to 'home' or other places through which they have made their migrant journeys. These practices work to make distinctions between an expatriate or not expatriate identity, with British migrants who have been abroad for longer periods of time being more likely to orient themselves away from the expatriate and look to articulate and claim different types of migrant belonging. This works to highlight two things. First, that there is a multiplicity of British migrant identities. Second, in exploring the sometimes contradictory ways in which British migrants understand what it means to be an expatriate in Singapore, we need also to appreciate the complexity and multiplicities of migrant identities.

This illustrates a need, in developing comparative research on British migration, to think more widely about the comparisons that we are making. This chapter then, like other work on expatriate identities, highlights that there is no singular expatriate experience (Fechter, 2007; Fechter & Walsh, 2010; Leonard, 2010; Scott, 2006; Walsh, 2006). This previous research on expatriates has highlighted the importance of understanding differences in experiences between gender, nationality, class, age and place (Beaverstock, 2005; Fechter, 2007; Yeoh & Willis, 2005). However, rather than explore this as being part of 'social fissures' (Scott, 2004, p. 392) within and between expatriate communities, the chapter highlights the importance of thinking about British migrants in relation to other British migrants to fully explore how migrant identities play out in place. Despite its often messy contradictions, viewing the expatriate as an orientation enables us clearly to see how the embodiment of expatriation can be interpreted as privilege within global regimes of power (Glick-Schiller & Salazar, 2013). As the chapter has shown, many of the practices that are associated with expatriates— brunches, travelling with ease around the region, clearly illustrate that being an expatriate is associated with opportunities that others do not have. We can see how the 'expatriate' can become seen as a marker of distinction between groups of British migrants, particularly by those who want to reject the term expatriate due to their feelings of the privileges it is associated with. But, the fact that they are able to reject the way in which their body is often marked by others illustrates privilege in itself. It shows that even for the British migrants who reject the term expatriate, privilege becomes part of their everyday lives. This I suggest is part of a practice of the British, and Western migrant more generally, in a Singaporean context, whether they understand themselves as being expatriate or not expatriate.

Notes

1 The respondent's names are pseudonyms to ensure the participant's anonymity.
2 The exception is the dependent pass. This visa is given to accompanying spouses and children to people given a work permit. You are not allowed to work on this visa and cannot carry out administrative practices such as opening a bank account.
3 The respondents were recruited via network events and snowball sampling.
4 To see discussions of how my respondents utilised race as a form of differentiation see Cranston (2017).

Acknowledgements

I would like to thank everyone who commented on earlier drafts of this chapter especially Dan Swanton and Allan Findlay, as well as my participants for assisting me with the research. The research was funded by the ESRC—ES/I018670/1.

References

Amit, V. 2007. 'Structures and dispositions of travel and movement'. In *Going First Class: New Approaches to Privileged Travel and Movement*, edited by V. Amit, 1–12. Oxford: Berghahn Books.

Beaverstock, J. 2005. 'Transnational elites in the city: British highly-skilled inter-company transferees in New York city's financial district'. *Journal of Ethnic and Migration Studies*, 31(2): 245–268.

Beaverstock, J. 2011. 'Servicing British expatriate "talent" in Singapore: exploring ordinary transnationalism and the role of the "expatriate" club'. *Journal of Ethnic and Migration Studies*, 37: 709–728.

Benson, M. & O'Reilly, K. 2016. 'From lifestyle migration to lifestyle in migration: categories, concepts and ways of thinking'. *Migration Studies*, 4(1): 20–37.

Berry, D. & Bell, M. 2011. '"Expatriates:" gender, race and class distinctions in international management'. *Gender, Work and Organization*, 19(1): 10–28.

Bourdieu, P. 1984. *Distinction: A Social Critique of the Judgement of Taste*. Cambridge, Mass: Harvard University Press.

Burrell, K. 2008. 'Materialising the border: spaces of mobility and material culture in migration from post-socialist Poland'. *Mobilities*, 3(3): 353–373.

Coles, A. & Walsh, K. 2010. 'From "trucial state" to "postcolonial" city? The imaginative geographies of British expatriates in Dubai'. *Journal of Ethnic and Migration Studies*, 36(8): 1317–1333.

Conradson, D. & Latham, A. 2005. 'Transnational urbanism: attending to everyday practices and mobilities'. *Journal of Ethnic and Migration Studies*, 31(2): 227–233.

Cranston, S. 2016a. 'Producing migrant encounter: learning to be a British expatriate in Singapore through the global mobility industry'. *Environment and Planning D: Society and Space*, 34(4): 655–671.

Cranston, S. 2016b. 'Imagining global work: producing understandings of difference in easy Asia'. *Geoforum*, 70: 60–68.

Cranston, S. 2017. 'Expatriate as a "good" migrant: thinking through skilled international migration categories'. *Population, Space and Place*. https://doi.org/10.1002/psp.2058.

Dunn, K. 2010. 'Embodied transnationalism: bodies in transnational spaces'. *Population, Space and Place*, 16(1): 1–9.

Fechter, A. 2007. *Transnational Lives: Expatriates in Indonesia*. Aldershot: Ashgate.

Fechter, A. 2010. 'Gender, empire, global capitalism: colonial and corporate expatriate wives'. *Journal of Ethnic and Migration Studies*, 36(8): 1279–1297.

Fechter, A. & Walsh, K. 2010. 'Examining "expatriate" continuities: postcolonial approaches to mobile professionals'. *Journal of Ethnic and Migration Studies*, 36(8): 1197–1210.

Finch, T., Andrew, H., & Latorre, M. 2010. *Global Brit: Making the Most of the British Diaspora*. IPPR.

Glick-Schiller, N. & Salazar, N. 2013. 'Regimes of mobility across the globe'. *Journal of Ethnic and Migration Studies*, 39(2): 183–200.

Ho, E. & Hatfield, M. 2010. 'Migration and everyday matters: sociality and materiality'. *Population, Space and Place*, 17(6): 707–713.

Hui, A. 2016. 'The boundaries of interdisciplinary fields: temporalities shaping the past and future of dialogue between migration and mobilities research'. *Mobilities*, 11(1): 66–82.

Knowles, C. & Harper, D. 2009. *Hong Kong: Migrant Lives, Landscapes, Journeys.* Chicago: University of Chicago Press.

Leonard, P. 2010. *Expatriate Identities in Postcolonial Organizations: Working Whiteness.* Farnham: Ashgate Publishing.

Ong, A. 1999. *Flexible Citizenship: The Cultural Logics of Transnationality.* London: Duke University Press.

O'Reilly, K. 2000. *The British on the Costa Del Sol: Transnational Identities and Local Communities.* London and New York: Routledge.

Schapendonk, J. & Steel, G. 2014. 'Following migrant trajectories: the im/mobility of sub-Saharan Africans en route to the European union'. *Annals of the Association of American Geographers*, 104(2): 262–270.

Scott, S. 2004. 'Transnational exchanges amongst skilled British migrants in Paris'. *Population, Space and Place*, 10(5): 391–410.

Scott, S. 2006. 'The social morphology of skilled migration: the case of the British middle class in Paris'. *Journal of Ethnic and Migration Studies*, 32(7): 1105–1129.

Urry, J. & Larsen, J. 2011. *The Tourist Gaze 3.0.* London: SAGE Publications.

van Bochove, M. & Engbersen, G. 2015. 'Beyond cosmopolitanism and expat bubbles: challenging dominant representations of knowledge workers and trailing spouses'. *Population, Space and Place*, 21(4): 295–309.

Walsh, K. 2006. '"Dad says I am tied to a shooting star!" Grounding (research on) British expatriate belonging'. *Area*, 38(3): 268–278.

Walsh, K. 2008. 'Travelling together? Work, intimacy and home amongst British expatriate couples in Dubai'. In: *Gender and Family among Transnational Professionals.* New York: Routledge.

Walsh, K. 2011. 'Emotion and migration: British transnationals in Dubai'. *Environment and Planning D: Society and Space*, 30(1): 43–59.

Yeoh, B. 2012. *Migration and 'Divercities': Challenges and Possibilities in Global-City Singapore.* [Online] Available at: www.kf.or.kr/file/pdf/Brenda%20S.A.%20Yeoh.pdf [Accessed 3 February 2016].

Yeoh, B. & Willis, K. 2005. 'Singaporean and British transmigrants in China and the cultural politics of "contact zones"'. *Journal of Ethnic and Migration Studies*, 31(2): 269–285.

Yeoh, B., Huang, S. & Willis, K. 2000. 'Global cities, transnational flows and gender dimensions, the view from Singapore'. *Tijdschrift voor Econmische en Sociale Geografie*, 91(2): 147–158.

5 Post-colonial liminality

'Expatriate' narratives in the East Africa Women's League

Sarah Kunz

Every Friday morning in Nairobi the 'East Africa Women's League' (EAWL) opens its doors for the weekly coffee morning and market. Situated in a secluded, leafy courtyard 'the League' was founded in 1917 and became British Kenya's key 'European'[1] women's organisation. One hundred years after its inception it still operates, although most members are past retirement age now and the League's previous prominence and influence have waned. Many League members, the majority of whom are still white and British, have lived in Kenya for most of their adult lives; many without acquiring citizenship. Leading inherently transnational lives, often stretched across the previous British Empire, they experience a conflicted sense of home. Based on ethnographic research in and around the EAWL, this chapter explores migrant narratives of belonging by tracing readings and negotiations of the category 'expatriate'. The term holds an important, yet not straightforward function in migrant discourses. It is specifically British migrants' narratives that I focus on in this chapter, although their white non-British peers largely shared in their constructions of belonging and elaborations of the 'expatriate'. In the few studies on Kenya's white minority, the term 'expatriate' is used to refer to temporary Euro-American residents, framed as a postcolonial phenomenon. Similarly, in much migration literature, 'expatriate' is taken, sometimes too self-evidently, to denote high-skilled, temporary labour migrants. While these studies commonly use 'expatriate' as an analytical category to describe a particular migrant group, this chapter explores the category as it is read and constituted by migrants themselves; 'expatriate' here emerges as a much more complex, ambiguous and troubled category.

As the term 'expatriate' circulates among EAWL members, their families and friends, it proves a malleable, instable category, met with both enthusiasm and aversion. Four main readings stand out: 'expatriate' as a short-term resident, a permanent resident, as non-citizenship status, and as post-colonial whiteness. These readings co-exist, often interchangeably employed by the same person. As argued in this chapter, the term 'expatriate' helps migrants belong partly through its very polysemy and ambiguity; contradictory uses are not mistakes but reflect the contradictions and tensions of belonging migrants expressed and reproduced. These were deeply entangled with the colonial past; the category 'expatriate' negotiated privileged liminal belonging as British migrants, geographically and temporally, straddled Kenya and the UK, the British Empire and its colony 'Keenya' (cf. Uusihakala, 1999:39). As such, 'expatriate' often folded colonial lifestyles, attachments and imaginations into post-colonial discourses without challenging fundamental structures of categorisation or their underlying racialised assumptions.

In the next two sections I review relevant literatures on white immigration to and belonging in Kenya and on transmigration, older-age and 'expatriate' identities. I then outline my methodological approach and introduce the EAWL, before tracing the 'expatriate' as it emerged in and around the EAWL. This discussion is structured around respondents' main readings and usages of the term: a first section explores seemingly contradictory readings of 'expatriate' as short-term versus permanent residence, a second section looks at its usage to articulate post-colonial whiteness, while a third section explores the constitutive absence of the term 'migrant' from respondents' narrations of 'expatriates' as non-citizens. In a concluding section, I then address more explicitly the term's tense polysemy and ambiguities and its role in narrating a sense of belonging that straddles colonial and post-colonial habits, attachments and discourses.

White migration and belonging in Kenya

British migration to Kenya formed part of the large-scale British emigration set in motion from the eighteenth century by domestic events and Britain's imperial politics (Kennedy, 1987). Kenya's white society was nationally diverse, but numerically dominated by immigrants of British 'stock', subsumed in a dominant 'British' culture and marked by a general sense of 'racial solidarity' (Kennedy, 1987:190). While the farmer-settler might dominate contemporary imaginations of Kenya's colonial society, the professional and business class 'consistently outnumbered white settler-farmers' (Wasserman, 1974:430–431); the colonial economy increasingly depended on commerce and industry and Nairobi became East Africa's commercial hub from the 1920s (Kennedy, 1987). Almost two-thirds of white people lived in urban areas by 1960 (Kyle, 1997). Also, colonial white society was marked by substantial transience: 'around half of all "settlers" were always transients, staying five years or less' (Lonsdale, 2010:75). Decolonisation brought ruptures and continuities. There was 'no mass exodus [of whites] before, during, or after independence' (Nardocchio-Jones, 2006:493)[2] and social change took hold only gradually. As my interlocutors suggest, white migrants that arrived to urban centres post-1963 independence found a in many ways, largely unchanged 'European' society.

A number of studies have examined white Kenyans' social position and belonging post-independence (Doro, 1979; Uusihakala, 1999; McIntosh, 2015, 2016; Fox, 2014). A recurring theme is anxiety regarding their position in Kenya, given their community's past role as colonisers and their continuing status as an economically and racially privileged minority. Key strategies to claim belonging in and importance to Kenya include performance of a deep relationship to the land and physical landscape, importance to national projects of development and conservation, and domestic projects of belonging through 'familial' relationships with black Kenyan staff (Fox, 2014; McIntosh, 2016). Yet, while the colonial past forms the backdrop of this scholarship, specific dynamics and strategies of translation from colonial to post-colonial modes of white belonging have received little attention. Moreover, scholarship has predominantly focused on white Kenyan citizens, particularly those with 'generational time-depth' and largely relies on a binary between white Kenyan citizens and 'expatriates' as 'temporary Euro-American visitors', framed as a postcolonial phenomenon (McIntosh, 2015:276–277). This does not reflect the diversity of migration and settlement experiences pre- and post-1963, nor take into account the complex ways 'expatriate' is used by long-settled migrants. As McIntosh (2015:277) notes, 'tens of thousands of whites […] immigrated to Kenya since its independence'. Some of these have been living there for decades without becoming citizens; likewise, not all 'settler'-descendants acquired citizenship at independence. Fox's (2014:356) question thus becomes particularly urgent for this group of largely British immigrants: 'what are the dynamics or political–economic structures enabling or enforcing them to maintain this conditional way of being?'.

Transmigration, older age and expatriate belonging

As now widely recognised, transmigrants forge specific forms of subjectivity and belonging as they create social, cultural and even political fields spanning two or multiple localities (Basch et al., 1994; cf. Gilmartin, 2008). Levitt and Glick-Schiller (2004) conceptualise transnational 'ways of belonging' as social relations and practices that enact a conscious connection to a particular group and identity. The challenge, they argue, is 'to explain the variation in the way that migrants manage that pivot' of belonging between assimilation and transnationalism (ibid:605). For this, it is crucial to recognise that migrants do not move in 'empty or geopolitically neutral space' (Grosfoguel et al., 2009:8). The wider historical, political and economic force fields that experiences of belonging are embedded in often remain insufficiently analysed (Ong, 1999; Glick-Schiller, 2005; Grosfoguel et al., 2009). The lives of British migrants that served, settled and laboured across the Empire depended on substantial social, political and economic transnational ties. This history shapes the transnationalism of especially older white British migrants in contemporary Kenya, which significantly differs from that of transmigrants less privileged by citizenship, class or 'race', more routinely studied in migration scholarship (Torres, 2006; Knowles & Harper, 2010; Amit, 2011; Croucher, 2012).

Analytical attention to post-colonial predicaments has marked more recent work on privileged and 'expatriate' migration (Fechter & Walsh, 2010; Amit, 2011). Much literature uses the category 'expatriate' to frame highly-skilled labour moving in international (labour) markets and de-localised networks (Beaverstock, 2002; van Bochove & Engbersen, 2013). Other work shows that even 'expatriate' migrants are always embedded in local hierarchies and histories and explores what Walsh (2010:235) calls the 'on-going traces of imperial power' in the lives of privileged migrants and the construction of 'expatriate' identities (Kothari, 2006; Leonard, 2008, 2010a, 2010b; Walsh, 2010, 2012; Fechter, 2010; Lester, 2012; Conway & Leonard, 2014). The centrality of whiteness for 'expatriate' migration is a recurring theme and scholars have explored its production and functioning in emotional, social and work spaces (Fechter 2005; Leonard 2008, 2010a, 2010b; Conway & Leonard, 2014; Lundström, 2014). Building on this work, but differing in analytical strategy, this chapter will 'follow the trajectories of the term itself' (Stoler, 2016:79), or follow 'expatriate' as a 'name-in-motion' (Tsing, 2015:29) to be explored in its own right: in moments of elaboration, in its contradictory and contested meanings, through its varied functions, silences and break-downs. As Tsing (ibid.) observes, 'if categories are unstable we must watch them emerge within encounters. To use category names should be a commitment to tracing the assemblages in which these categories gain a momentary hold'. So, rather than relying on 'expatriate' to frame a particular group of migrants, this chapter studies the term's elaboration in a particular social space. This chapter contributes in another way; few studies have explored how those considered 'expatriates' negotiate their status *as migrants*, and view the relationship between the two labels 'expatriate' and 'migrant' (although see Rogaly & Taylor, 2010; Walsh, 2012, 2016).

While there is increasing scholarship on older migrants (Walsh & Näre, 2016), literature exploring 'expatriate' identities often focuses on younger (labour) migrants. An exception is work on 'lifestyle' or 'retirement migration', studying Northern Europeans or North Americans residing in the Mediterranean or Central America (O'Reilly, 2000; Ciobanu et al., 2017). Yet, despite similarities, 'ageing as an immigrant is not the same as migrating as an elder' (Torres, 2006:1352). And if 'ageing immigrants' have so far received little academic attention, as Zontini et al. (2015) claim, aged immigrants privileged along dimensions of class, nationality or 'race' have received even less (Walsh & Näre, 2016). Thus, the British migrants foregrounded in this chapter, who 'aged in migration' and occupy privileged positions of class, race and citizenship find little equivalent in literatures on older migrants or 'expatriate' identities. Their negotiations, appropriations and rejections of the term 'expatriate' shed light on how older migrants negotiate belonging, identity and privilege within wider transnational histories of power and specific sites of post-colonial tension.

Placing migrants in 'the League'

This chapter orbits around the 'East Africa Women's League' (EAWL), or 'The League', where I conducted ethnographic and archival research between May and October 2016. Founded 1917 in Nairobi with membership restricted to 'European', i.e. white women, and often prominent leaders of the British colonial elite, the League gained political and social influence. After attaining the vote for Kenya's white women, its aim was 'to study and take action on, where necessary, all matters affecting the welfare and happiness of women and children of all races in East Africa' (van Tol, 2015:433). Despite later claiming an 'apolitical' character, it was 'a key outlet for the public partici-pation of white women in Kenya' (ibid:435; cf. Anderson, 2010). The League promoted, and policed, middle-class virtues of female domesticity, family, and philanthropy, key to constructions of British imperial white womanhood and thus the colonial project. League membership changed with independence as many 'Europeans' left Kenya and membership was begrudgingly opened to 'Asian' and 'African' women. Yet, membership remained white by a vast majority, now both Kenyan citizens and non-citizens. That non-citizens remained a significant group is suggested by recurring anxieties about immi-gration legislation in League council meetings post-1963. Similarly, falling membership and a lack of younger members caused continued concern and reflected failure or unwillingness to recruit non-white members. Nowadays, the League acts as a central social space for older members that are still majority white British; and as in colonial days, non-British members share, at least at the League, in an overarching cultural 'Britishness'.

This chapter grows out of research exploring the historical and con-temporary constitution of 'expatriate' identities in Nairobi. Situated in this wider project, it specifically draws on five months of ethnographic research conducted at the EAWL's Nairobi Headquarters, during which I inquired into migrant narratives of belonging and, specifically, traced the category 'expatriate' among this older migrant cohort. It also draws on five expert interviews on Kenyan immigration law and practice with migration research-ers and practitioners, an immigration lawyer and an immigration agent and former government official. My ethnographic research at the League included participant observation, archival research, 13 life history interviews with League members, family and friends, and countless informal conversations. I took part in the League's weekly coffee mornings and other social and chari-table activities in and outside the League, including branch meetings. Reflecting League membership, most interlocutors I met there were white British, in their sixties to eighties, married in heterosexual relationships or widowed, and had lived in Kenya since the 1950s to 1970s. Most were retired and many women had been housewives, though some had worked before, during or after their marriages. More specifically, of 13 respondents (9 women, 4 men) in 11 life history interviews, 12 self-identified as white; 9 were British, 2 Western European, and 2 naturalised Kenyans (previously Indian and British); one was born in colonial Kenya but remained a British citizen, the 'newest' arrived in Kenya 17 years ago, most others arrived in the 1950s or 1960s. Respondents had worked, for instance, as teachers and secretaries, in the tea industry, the late British colonial administration and later NGOs. Interviews lasted between one and five hours and generally took place at respondents' homes. It is specifically British migrants' narratives that I focus on in this chapter, although white non-Brits largely shared in their discourses of belonging and constructions of the 'expatriate'.

'Expatriate' narratives in the League

In the following sections I trace the category 'expatriate' as a lynchpin in older white British migrants' narrations of belonging. This discussion is structured by respondents' readings and usages of the term. I first explore seemingly contradictory readings of the 'expatriate' as short-term versus permanent residence, I then discuss its usage to articulate post-colonial whiteness. Third, I explore the 'expatriate' as a non-citizen, arguing that it does work to exclude the category 'migrant' from discourses of belonging. The concluding section then addresses more explicitly the tense polysemy and ambiguities of the term 'expatriate'.

'Two-Year-Wonders', 'permanents' and modes of belonging

'Expatriate' was a term fervently rejected by some League members on the grounds that it referred to those who only lived in Kenya temporarily.[3] On my first day at the League, I sat and chatted with some women about my research. Sally and her husband came to Kenya roughly 30 years ago from the UK, 'all us expats ...', she casually says, referring to League members. Opposite her, I can see Corinne hesitate. Just moments before she told me that she did not feel like an expat (any more). She interrupts Sally: 'I don't really consider myself an expat though'. Sally turns to me and explains that 'expats weren't liked', they were called 'Two-Year-Wonders' (fieldnotes). This brief episode reveals the ambiguous status of the category 'expatriate'. Sally first used 'expatriate' as a vague 'us', looking over a room of older, mainly white women of British origin or citizenship settled in Kenya for decades – and then equated it with temporary residents.

'Two-Year-Wonders' was used pervasively and always somewhat pejoratively, even mockingly, corresponding to how McIntosh (2015:254) encountered the term, used for 'merely transient' 'Euro-American expatriates'. However, while some League members might have arrived on short-term contracts, I did not encounter any 'Two-Year-Wonders' there. The figure remained important as one against which members, whether Kenyan citizens or not, could establish their own belonging in Kenya: by comparing themselves to 'whites still more "foreign," still more Other to Kenya' (ibid:264). This practice in fact has a colonial genealogy in which the League itself is entangled. As in other colonial contexts (Leonard, 2008), colonial society was marked by substantial transience (Lonsdale, 2010) and deriding 'new immigrants' and their 'ignorance of colonial conditions' was common among Kenya's white community (Kennedy, 1987:141). The League, too, positioned itself as essential to the colony against 'newcomers' and through 'educating' them (Anderson, 2010). The 'expatriate' as temporary resident is thus inserted in techniques of belonging that grow out of colonial practices.

However, usage of 'expatriate' was more complicated than a simple equation with 'temporary residents'. Indeed, others used 'expatriate' to refer to permanent residents. As Cedric told me while sipping Gin and Tonic on his shaded veranda:

> there is a very big population of expatriates–, what do you call aid workers, same thing? Because they are not permanent. Because expatriate, well I think of it as people who have been here a long time, and are more or less permanent.

Cedric was in his early eighties and while he had spent more or less all his adult live in East Africa and the majority of it in Kenya, he had remained a British citizen. He did not want to return to the UK, he told me the climate did not agree with him and he enjoyed life in Kenya, where he owned some land. In a similar use of 'expatriate', Isabel told me about her husband Michael, who had lived in Kenya since the 1960s: 'He is *the* quintessential expatriate, I'd say. He'll never leave Kenya' (fieldnotes). 'Expatriate' in such instances described subjects that imagined themselves simultaneously as British in terms of nationality and culture and as at home in Kenya; as belonging in – but not necessarily to – Kenya.

In fact, across and below statements about temporality, more nuanced assessments of what made 'expatriates' emerged that hinged on lifestyles, attachments and commitments: on transnational 'ways of belonging' (Levitt & Glick-Schiller, 2004). Ultimately, the signifier 'expatriate' was often an assertion about practices, relationships and sentiments connected to how and where one belonged. Helen expressed 'I want to remain sort of British' and Cedric explained being an 'expatriate' as 'I suppose one values one's sort of background and way of life and tradition and all that sort of thing'. While most had no desire to live in the UK, they travelled there regularly and extensively, sent their children for schooling or university and retained what they viewed as 'British' cultural habits and traditions, like playing bridge or proudly identifying as monarchists. 'Expatriate' denoted ways of being and belonging that hinged on the desire and entitlement to live in Kenya on one's own terms, including re-producing an, at times, clichéd classed 'British way of life'. A critical assessment was offered by Dorothea:

> Expats are the people who have come here, mainly the English because this was an English colony[4] after all, who brought with them all their

heritage, their ideas and everything and that have held onto it. And they stay here, but continue to live their lives like they would at home. And they just enjoy the climate here, and certain other things.

Like others, Mary expressed insecurity about what exactly 'expatriate' meant. First referring to temporary residents, she subsequently qualified her explanation:

> I think there are those people who will always consider themselves expatriate, based on where they have come from, it doesn't matter how long they will be here. There are others who have embraced, to some extent, the Kenyan culture. [...] So, expatriates, when I hear that word I just think of colonialism.

In these assessments, 'temporariness', rather than describing an actual time-period, described a form of belonging that hinged on *performed* temporariness through continued association with another 'Home' (Walsh, 2016). For both Dorothea and Mary 'expatriate' denoted ways of life that reproduced modes of being and belonging from colonial 'Keenya': 'the central feature of the settler culture was its renunciation and repression of any substantive adaptation to the host environment' (Kennedy, 1987:189). In fact, 'colonial settlers [...] kept Europe as their myth of origin and as a signifier of superiority even when formal political ties and/or dependency with European colonial powers had been abandoned' (Stasiulis & Yuval-Davis, 1995:5). If British settlers belonged *in* Kenya precisely by belonging *to* Britain, then this ultimately justified that Kenya belonged to them. Variably positioning oneself as part of Kenya *and* the UK, as here encapsulated by the 'expatriate', emerges as a modality and strategy of belonging echoing British colonial lives; as such, it ultimately encompasses claims to power and privilege.

Notably, Mary and Dorothea had come to Kenya married to 'Asian' and 'African' Kenyans, whom they had met while their prospective husbands pursued further education in the UK. Among white League members, it was women like them that, from their 'marginal positioning' (Leonard, 2008:54), most criticised 'expatriate' modes of belonging, contrasting it to their own integration and efforts to adjust to their husbands' cultures. Older women like Dorothea and Mary would recount early years in Kenya, when they would 'only be with their husband's community'. Tellingly, they all joined the League relatively recently, as they explained, because of earlier pre-occupations with family and work lives. Yet, as an Asian Kenyan League member noted dryly, they would not have been very welcome at the League in early post-independence decades, having transgressed colonialism's central racial boundary in their marriages (Kennedy, 1987). As older women they joined the League mainly for its social life; their husbands had passed away in the meantime, possibly further facilitating their reconnecting with the white British community. While some privately critiqued what they perceived to be its on-going colonial character and narrated themselves as 'not really part' of the League, and unlike 'expatriates', they shared many of their peers' privileges, and to some extent relied on the League socially and emotionally, partaking in its reproductions of a 'white Britishness'.

Post-colonial whiteness in limbo

Whiteness was central to white migrant discourses; belonging in its various modes always meant a belonging conditioned by whiteness. 'Expatriate' itself was regularly used to denote 'whiteness', as became apparent in a conversation with Michael about Muthaiga Country Club[5] (fieldnotes):

SK: So how many of the members would you say are expatriates?
MICHAEL: I would say we make up around 30 percent now.
SK: And who are the other 70 per cent?
MICHAEL: Well, the Asians and the Africans.

'Expatriate' is here used synonymously with the colonial 'European', against the categories 'African' and 'Asian'. Used like this, 'expatriate' did not distinguish citizen from non-citizen, short from long-term residents. It implied whiteness as a shared 'European' ancestry and on-going community – with unclear status in independent Kenya. As the very presence of 'African' and 'Asian' members at Muthaiga Club or the League testified, as an instrument of power whiteness had been fundamentally challenged. Whether historically accurate or not, interlocutors narrated 'expatriate' as a term that gained traction post-independence. Camilla told me that 'expatriates' had been around before, mainly 'farmers', but also 'white telephonists' and 'taxi drivers', however, they were not called 'expatriates' 'because it was before independence – and I suppose most of them thought they were there forever, and it was their land and so on'. In contemporary imaginations of the colonial past, the belonging and role of white people in Kenya was secured and self-evident. The term 'expatriate' was needed not because a new group of people had arrived or an altogether new subjectivity developed; rather, it narrated changed external conditions. Whiteness continued to be a central dimension of older white migrants' self-identifications, but as white people they felt in 'limbo' in contemporary Kenya. As Conway and Leonard (2014:2) note for South Africa, 'being "white" is still very much part of the grammar of South Africa and remains a key marker of identity, even though it "just isn't what it used to be"'.

As such, the term 'expatriate' did work to fold colonial imaginations of whiteness, its role and constitutive relationships into new language, whilst often re-producing their content. The meaning and role of whiteness was still habitually constructed against an 'African Other' imagined as staff or potential recipients of charity. As Dorothea told me about a friend:

She is a real expatriate, has been here for ages, but sees herself as a consultant. They call her Mama Africa, she owns a camp in the Maasai Mara, has worked with the Maasai, taught them about condoms and all. But throughout she has kept her Englishness, still looks over there.

Framed by a narrative in which the 'English' woman guides and directs 'African' learning and progress, this not uncritical mobilisation of 'expatriate' as 'consultant' reconciles colonial imaginations of the social order structured by inequalities in knowledge, perspective and matters of 'civilization', couched in romanticised paternalistic benevolence. Like here, the term 'expatriate' often described social relationships steeped in colonial histories and imaginations; yet, without necessarily confronting their imperial roots and presents.

The Friday coffee mornings at the League, which branches took turns organising, seemed a lynchpin in reproductions of white sociality. As Leonard (2010a:109) discusses for Hong Kong, 'space and place are integral to the production and performance' of white identities. Similarly, the League itself continued to be imagined – at least by white members – as an essentially white, British organisation and space. While the few Asian Kenyan and some white League members mixed socially, boundaries between 'whites' and 'Africans' seemed largely maintained in everyday social interactions and imaginations. Social segregation along inherited boundaries tracked with a view shared at least by some whites that, as Jean put it, 'in the old days everyone stayed within their own communities, but that's how everyone liked it'. This historical understanding attests to 'ignorance' 'as a social achievement with strategic value' in constructions of whiteness (Steyn, 2012:8). It also involved a claim on the present as many members, whether consciously or out of habit, reproduced colonial Kenya's socially segregated society of 'races that meet in the marketplace but do not mingle' (Doro, 1979:45). Accordingly, Jim, an older Brit, once looked around the hall, predominantly occupied by older, white women and remarked: 'there were many expats that left at independence, but I think the ones that stayed were actually quite happy' (fieldnotes). Within and through the League, white members defended not only their interpretation of what it meant to be white but also a desire to re-create social space on their own terms.

The handful of 'African ladies' that attended coffee mornings accordingly proved an uncomfortable presence, but also afforded opportunities for reproducing the value and shared identity of whiteness. While the 'African ladies' generally kept to themselves and were not talked to, they were regularly talked about, with a mix of unease, resentment, and benevolent paternalism. Complaints I heard about them often reproduced normative classed, white femininity. I was told that they were too loud or put no effort into organising coffee mornings:

DOROTHEA: Our women, they can bake–
SK: When you say 'ours' you mean your branch?

DOROTHEA: No, I mean our Europeans. [...] For the Africans, this is a duty.

Corinne told me 'even now we still have some members who have a very colonial attitude, which I am working very hard to get rid of. [...] They are very patronising about our African members particularly–'. However, instead of describing instances of paternalism, she continued: '–there is one group of African ladies, and they are really nice ladies. They are working very hard to make sure that they comply, that they do what we are about, but they don't really understand a lot of what we are about'. The 'we' of the organisation is here discursively constructed against compliant, yet not quite capable 'African' members and the League thereby reproduced as an essentially white organisation. Even women who critiqued the League's 'colonial character', often participated in reproducing a space organised along racial lines and a whiteness imagined in colonial ways.

It was also Corinne who told me about 'African expats', explaining: 'see, you also got expat Africans. [...] Because, you know, you are looking at expats as Europeans. But there are a lot of Africans who could be considered as expats'. In fact, I was not focusing on 'Europeans'. This assumption rather reflected Corinne's own and her peers' habitual deployment of the term 'expatriate'. The very expression that Africans 'could' be considered expats highlights that in this space they were usually not; and the adjectival description of the '*African* expat' betrayed how the 'expatriate' was ordinarily tethered to whiteness (no respondent would specify '*European* expat'). Accordingly, when I mentioned to Marlene that Corinne had introduced me to 'African expats', she seemed surprised, almost offended, that 'I'd call them expats, too' (fieldnotes). On yet another occasion, one of the few younger members from India exclaimed she was 'very much' an 'expat': 'we have lived all over Africa for 17 years, and have now been here for four years'. It was older, white migrants that assumed or actively claimed the term 'expatriate' as 'whiteness', mobilising it to translate inherited notions of white privilege into new discourses, frequently without fundamentally challenging the basic structure of categorisation or its underlying assumptions.

Expatriates, migrants: Legal liminality and definitional struggles

'Expatriate' was also used to reference non-citizenship status, taking up a discursive space that could have been filled by the category '(im)migrant' – but with different connotations. For decades, many respondents had been reluctant to 'trade' British or other Northern citizenship for Kenyan citizenship and consciously transferred their citizenship to their children. This was principally because of the ease of international travel and a sense of security their passports provided – particularly given Kenya's perceived uncertain future and their place within it (Doro, 1979). Migrants thus lived in Kenya on a variety of passes and permits, including retirement passes, permanent residence and work permits. Many, women especially, were legal 'dependents' of partners or children with (dual) Kenyan citizenship. Others, I was told, had lived and worked irregularly on visitor passes. Non-citizens also included Brits born in pre-independence Kenya who had not registered as Kenyan citizens in the post-independence two-year window. Many thus consciously practiced strategies of long-term 'legal liminality' (Menjívar, 2006) and a legal, social and cultural 'flexibility in geographical and social positioning' (Ong, 1999:3). However, strategies of legal and social liminality were largely rooted in privilege rather than exclusion.

The term 'expatriate' was often used to denote non-citizens, who did not view themselves, nor were they viewed by their peers, as '(im)migrants' – a seemingly unrelated category (Rogaly & Taylor, 2010; Walsh, 2016). The 'categorical' boundary maintenance respondents engaged in sometimes resulted in complicated manoeuvres of argumentation:

SK: Do you ever think of yourself as an immigrant here?

MICHAEL: Oh no, no, no never. I've always had choices to return; I think if you're an immigrant, you, that is to some extent a very final choice, isn't it? And maybe it's not a choice, again, for you. You're obliged to go, to survive. Well no. That wouldn't apply to me.

Michael here relies on the element of 'choice' to dissociate himself from the 'immigrant'. Yet, just a moment previously he had included Brits without a 'choice to return' as a type of 'expatriate':

MICHAEL: There are many types of expatriates [...] maybe there are people here that had to stay on [at independence]. That had nowhere to go to in the UK, they didn't have the money, they didn't have the connections, they didn't have the skills. And they couldn't go anywhere. They had to stay. Yes, so that's a different kind of expatriate altogether.

Michael's narration is self-contradicting in its attempt to disambiguate (British) 'expatriates' from 'immigrants'. Similarly, Marlene seemed bewildered when asked if she ever considered herself an immigrant in Kenya, where she had lived since the 1960s and intended to remain: 'No I'm not an immigrant, I came here because my husband got offered a job' (fieldnotes).

In fact, the eclipse of the category '(im)migrant' from discourses of belonging seemed part of the performative achievement of the category 'expatriate'. Narrating oneself as an 'expatriate' represented a claim to mobility and immobility on one's own terms and resistance to the actual or potential challenges to autonomous (im)mobility associated with the category '(im)migrant'. In this context, dealings with immigration matters emerged as moments of 'categorical' power struggles, where self-definition did not always match other-definition. As a Kenyan refugee advocate and immigration law expert I interviewed joked: 'these days we make them ['expatriates'] feel as migrants'. He continued: 'but it's not a Kenyan phenomenon alone. Tanzania has enforced it, and a lot of Kenyans have been forced back home'. This statement not only challenged the privileged border treatment associated with 'expatriates', it broke open the frequently racialised boundary separating the categories 'expatriate' and 'migrant' by placing 'expatriates' in Kenya and Kenyans abroad in the same analytical framework, as migrants. This was the very boundary respondents strove to uphold in efforts to sustain inherited rights to live in Kenya, individually and as a white community, largely on their own terms. Such entitlement was destabilised by association with the figure 'migrant' or comparison with those whom respondents traditionally labelled 'Africans'; especially when some respondents complained, like Marlene, that there was 'too much immigration in Europe' and that immigrants 'don't integrate' (fieldnotes).

Yet, many non-citizen League members worried about their legal statuses, especially given older age's limited mobility, deteriorating health and increased vulnerability. Different immigration statuses implied varying degrees of stability and a range of experiences with the immigration bureaucracy. Yet, respondents did not understand negative experiences as those of older immigrants with insecure legal statuses navigating a potentially inefficient and sometimes corrupt, but also largely lenient immigration apparatus. Instead, they felt collectively targeted as a 'white community'. Marlene, upon rejecting the label of immigrant, talked about a British friend, whose citizenship application had been rejected: 'She has lived here all her life and doesn't know anywhere else, they are in their 80s now, [...] isn't that the most horrible example of racism in reverse' (fieldnotes). Another Friday morning, Jessica told me her life story; upon Dorothea introducing her as an expat, Jessica hesitated over whether she was an expatriate, 'well', she then shrugged, 'I don't have citizenship'. She and her husband had lived in Kenya on and off since the 1960s. Now he was on a retirement pass and she his legal dependant. They wanted to stay but felt their lives to be 'in limbo': 'it's horrible, really how they are treating us. I know some people are trying to get citizenship but aren't getting it. They are trying to push us out really' (fieldnotes). Individual anxieties were real and sometimes existential. Yet, their articulation as assaults on a white community mostly evidenced respondents' own pre-occupation with whiteness, tapping into what Anderson (2010:84) calls 'a deep well of anxiety about their vulnerability amid a hostile African majority'; a majority which appeared disinterred, rather than hostile.

This collective anxiety not only glossed over positive experiences, it constructed whiteness as a specific hurdle in dealing with immigration. Kenyan immigration and citizenship law has been critiqued for discrimination against women, refugees and numerous long-resident stateless populations, like the Kenyan Nubians, Kenyan Somalis, and Coastal Arabs (Mucai-Kattambo et al., 1995; Open Society Justice Initiative, 2011). Unclear processes and corruption have been further challenges to an effective and fair operation of the law; and public debate largely frames migration as a security issue related to the figure of the 'refugee' (expert interviews). Those racialised as white, especially with passports from the Global North have, if anything, likely benefited from assumptions about their status and weak implementation of legal requirements. Tellingly, 'legal liminality' was for some predicated on temporary or permanent irregularity, residing and working on visitor passes. Thus, while white Kenyans' social status may be contested, as McIntosh (2015, 2016) argues, white *immigrants* at least appear privileged vis-à-vis other immigrant, refugee and stateless populations. While enforcement of immigration regulation was becoming stricter, immigration experts generally agreed that: 'when you say people are finding it difficult now, I would say yes and no. Yes to the extent that people were used to a system that wasn't working, a system that wasn't strict' (interview, immigration agent). Especially Brits seemed to have inherited a sense of entitlement of being in Kenya that prevented many from anticipating or preparing for challenges they now faced. As Walsh (2012:57) argues, 'many Britons carry with them an understanding of their migrant status which fails to match the contemporary world'.

Malleable categories, ambiguous belonging and post-colonial liminality

This chapter discusses the belonging of older white British transmigrants associated with the East Africa Women's League through tracing 'their' category 'expatriate' – its polysemy, ambiguities and tensions. 'Expatriate' narrated modes of being and belonging that straddled the colonial and postcolonial, resulting in what can be described as liminal post-colonial belonging. Turner (1967:95–96) argues that a 'liminal persona' emerges in transitional periods, 'at once no longer classified and not yet classified', 'neither one thing or another; or maybe both'. Migration scholarship has employed the concept of liminality largely to explore experiences of legal and political exclusion and marginalisation, showing that '*extended* "legal liminality"' can arise, including indefinite 'grey areas between documented and undocumented' (Menjívar, 2006:1004). Yet, liminality does not necessarily have to originate in powerlessness versus state apparatuses. Wang (2016:1942) argued that Chinese Americans 'use their liminality or "strategic in-betweenness"' as they 'leverage Western training with their assumed knowledge of Chinese culture to create personal economic advantage'. Likewise, liminality can here be seen as a migrant 'strategy', hinging on transnational lifestyles and attachments that are decidedly post-colonial – neither a simple replica of the colonial, nor having transcended colonial modes of being, emotionally, culturally or socially.

This post-colonial liminal belonging was an achieved state, to some extent actively reproduced; and although not uncontested, conditioned by relative power and associated privileges that especially British migrants had come to expect. It was to some extent migrants' unwillingness to re-define the terms of their being and belonging that produced their liminality – legally as much as emotionally. Yet, the *desire* to maintain transnational ties and hold onto 'one's way of life' is maybe not what distinguishes respondents from other transmigrants. What ultimately seemed to make 'expatriates' was the ability to do so, temporarily or permanently, and largely on their own, privileged terms. This had, as recognised by some, a colonial genealogy; but ultimately depended as much on a particular colonial history as on the continuing political and economic power of the Global North. Older British migrant lives and attachments rested in (and on) the on-going coloniality of power. Being an 'expatriate' generally relied on access to specific forms of transnationalism afforded by Northern citizenship, education, cultural capital, and social networks. Yet, this structural position of privilege ultimately cut across self-identifications as 'expatriate'; most British migrants had, as McIntosh (2016:29) put it, 'kept an exit door open'. Possibly, some migrants needed to perform belonging in and commitment to Kenya by deriding 'Two Year Wonders', also because they were not fully convinced of their own commitment.

McIntosh (2015:273) writes that 'white Kenyans aren't wrong that the discourse framing them as "settlers" and interlopers is essentialist'. Yet, colonial imaginations of whiteness were not only imposed on, they were also re-produced by white residents. Respondents did not suggest that 'ethnic and racial differences aren't relevant to belonging' (ibid.); it seemed, rather, they wanted to decide *when* they were relevant and *how*. While whiteness was central to migrant belonging, it was a whiteness in 'limbo', aware of its relative loss of power. As such, the League itself was a liminal space that white Brits could inhabit largely on their own terms, if without their prior centrality and importance. 'Expatriate' was instrumental in narrating this whiteness in limbo, partly through its very polysemic ambiguity. Further, as Stoler (2016:70) highlights, associations as much as dissociations are productive of meaning and concepts do not only reflect power but 'intervene in the allocation of power'. The presence of the 'expatriate' was related to the discursive absence of the 'migrant' – a substitution complicated by *immigration* difficulties. Being treated as 'migrants' challenged post-colonial liminal belonging predicated on the entitlement to be in Kenya on one's own terms, suggested vulnerability and sparked collective anxiety. Yet, interpretations of immigration difficulties as an assault on whiteness relied on an imagination of whiteness as central to Kenyan society and politics, an assumption that itself seemed like a distinctly 'white phenomenon'; especially given its reliance on a homogenised 'African' failed to note divisions and discriminations among 'Africans' similarly playing out in the immigration apparatus.

This chapter is not guided by the search for a correct or original meaning of 'expatriate'. Indeed, a key insight is that 'expatriate' is a polysemic, ambiguous and malleable category; however, these are not random and they do emotional and political work. Migrants' material and narrative responses to their historical and on-going privileges are fragmented, frequently evasive and at times self-contradictory (Leonard, 2008:57). It was the experience and also strategy of remaining emotionally, conceptually and at times legally in limbo that the term 'expatriate' helped negotiate. Contradictions in usage are thus not 'mistakes' and ambiguity remains unresolved, as did the tensions of belonging it speaks to (Doro, 1979; Uusihakala, 1999). However, this is not to say the category 'expatriate' is 'innocent' or infinitely stretchable. It consistently carried connotations of whiteness and privilege, as confirmed by a range of scholarship (Fechter, 2007; Walsh, 2010; Leonard, 2010a). Not acknowledging the violent roots of social positions and imaginations was at the core of migrants' simultaneous (non)belonging in Kenya and required 'categorical' diversionary tactics; ambiguity and a constant 'evasive turning away' (Stoler, 2016:255). Ultimately, migrants' practices and narratives of belonging represented personal as much as political projects that sometimes re-created colonial imaginations and relations, at other times tried to transcend or found themselves pushed beyond them. They conveyed inherited privileges and newer anxieties. Ultimately, they also complicate neat temporal classifications like the colonial and postcolonial, past empire and present globalisation.

Notes

1 Throughout this chapter, 'European' denotes whiteness (positioned against 'Asian' and 'African'). This represents its colonial usage and echoes how respondents still used the term now.
2 Whites were estimated at 53,000 in 1963 and 40,000 in 1978 (Doro, 1979). Kenya's 1989 census still counted 3,184 white citizens and roughly 31,000 other whites (Uusihakala, 1999).
3 Also for methodological reasons, I tried not to 'label' anyone 'expatriate'; I communicated that my research was an inquiry into belonging and specifically into what 'expatriate' meant and 'expatriate lives' entailed; a project for which I was seeking both 'expatriate' and non-'expatriate' contributions, leaving it up to respondents to decide which group they belonged to.
4 Dorothea is not British and the interview was conducted in her native language, in which the distinction between 'English' and 'British' does not carry much weight. This might explain why she frequently failed to distinguish the two.
5 Once (in)famous as Kenya's white setter club, Muthaiga Country Club now hosts Kenyan and foreign elites.

Bibliography

Amit, V. ed., 2011. *Going First Class? New Approaches to Privileged Travel and Movement*. Oxford: Berghahn Books.
Anderson, D.M. 2010. 'Sexual threat and settler society: 'Black perils' in Kenya, c. 1907–1930'. *The Journal of Imperial and Commonwealth History*, 38(1): 47–74.
Basch, L., Schiller, N.G. & Blanc, C.S. 1994. *Nations Unbound: Transnational Projects, Postcolonial Predicaments and Deterritorialized Nation-States*. London andNew York: Routledge.
Beaverstock, J.V. 2002. 'Transnational elites in global cities: British expatriates in Singapore's financial district'. *Geoforum*, 33(4): 525–538.
Ciobanu, R.O., Fokkema, T. & Nedelcu, M. 2017. 'Ageing as a migrant: vulnerabilities, agency and policy implications'. *Journal of Ethnic and Migration Studies*, 43(2): 164–181.

Conway, D. and Leonard, P. 2014. *Migration, Space and Transnational Identities: The British in South Africa.* Houndmills, Basingstoke: Palgrave Macmillan.

Croucher, S. 2012. 'Privileged mobility in an age of globality'. *Societies*, 2(1): 1–13.

Doro, M.E. 1979. '"Human souvenirs of another era": Europeans in post-Kenyatta Kenya'. *Africa Today*, 26(3): 43–54.

Fechter, A.-M. 2005. 'The "Other" stares back: experiencing whiteness in Jakarta'. *Ethnography*, 6(1): 87–103. https://doi.org/10.1177/1466138105055662.

Fechter, A.-M. 2007. *Transnational Lives: Expatriates in Indonesia.* Aldershot, Hants, England and Burlington, VT: Ashgate.

Fechter, A.-M. 2010. 'Gender, empire, global capitalism: colonial and corporate expatriate wives'. *Journal of Ethnic and Migration Studies*, 36(8): 1279–1297.

Fechter, A.-M. & Walsh, K. 2010. 'Examining "expatriate" continuities: postcolonial approaches to mobile professionals'. *Journal of Ethnic and Migration Studies*, 36(8): 1197–1210.

Fox, G. 2014. 'Commitment issues: security and belonging in a white Kenyan household'. *Canadian Journal of African Studies*, 48(2): 353–372.

Gilmartin, M. 2008. 'Migration, identity and belonging'. *Geography Compass*, 2(6): 1837–1852.

Glick-Schiller, N. 2005. 'Transnational social fields and imperialism: bringing a theory of power to transnational studies'. *Anthropological Theory*, 5(4): 439–461.

Glick-Schiller, N. & Levitt, P. 2004. 'Conceptualizing simultaneity: a transnational social field perspective on society'. *International Migration Review*, 38(145): 595–629.

Grosfoguel, R., Cervantes-Rodriguez, M. & Mielants, E. 2009. *Caribbean Migration to Western Europe and the United States: Essays on Incorporation, Identity, and Citizenship.* Philadelphia: Temple University Press.

Kennedy, D.K. 1987. *Islands of White: Settler Society and Culture in Kenya and Southern Rhodesia, 1890–1939.* Durham: Duke University Press.

Knowles, C. & Harper, D. 2010. *Hong Kong: Migrant Lives, Landscapes, and Journeys.* Chicago: University of Chicago Press.

Kothari, U. 2006. 'Spatial practices and imaginaries: experiences of colonial officers and development professionals'. *Singapore Journal of Tropical Geography*, 27(3): 235–253.

Kyle, K. 1997. 'The politics of the independence of Kenya'. *Contemporary British History*, 11(4): 42–65.

Leonard, P. 2008. 'Migrating identities: gender, whiteness and Britishness in postcolonial Hong Kong'. *Gender, Place and Culture*, 15(1): 45–60.

Leonard, P. 2010a. *Expatriate Identities in Postcolonial Organizations.* Farnham: Ashgate.

Leonard, P. 2010b. 'Work, identity and change? Post/colonial encounters in Hong Kong'. *Journal of Ethnic and Migration Studies*, 36(8): 1247–1263.

Lester, A. 2012. 'Foreword'. In *The New Expatriates: Postcolonial Approaches to Mobile Professionals*, edited by A.-M. Fechter & K. Walsh, 1–8. London: Routledge.

Levitt, P. & Glick-Schiller, N. 2004. 'Conceptualizing simultaneity: a transnational social field perspective on society'. *International Migration Review*, 38(3): 1002–1039.

Lonsdale, J. 2010. 'Kenya: home county and African frontier'. In *Settlers and Expatriates: Britons Over the Seas*, edited by R. Bickers, 75–111. Oxford and New York: Oxford University Press.

Lundström, C. 2014. *White Migrations: Gender, Whiteness and Privilege in Transnational Migration*. New York: Palgrave Macmillan.

McIntosh, J. 2015. 'Autochthony and "family": the politics of kinship in white Kenyan bids to belong'. *Anthropological Quarterly*, 88(2): 251–280.

McIntosh, J. 2016. *Unsettled: Denial and Belonging Among White Kenyans*. Oakland: University of California Press.

Menjívar, C. 2006. 'Liminal legality: Salvadoran and Guatemalan immigrants' lives in the United States'. *American Journal of Sociology*, 111(4): 999–1037.

Mucai-Kattambo, V.W., Kabeberi-Macharia, J. & Kameri-Mbote, P. 1995. 'Law and the Status of Women in Kenya'. In *Women, Laws, Customs and Practices in East Africa – Laying the Foundation*, edited by J. Kabeberi-Macharia. Geneva: International Environmental Law Research Centre. Available at: www.ielrc.org/content/a 9501.pdf [Accessed 4 September 2017].

Nardocchio-Jones, G. 2006. 'From Mau Mau to Middlesex? The fate of Europeans in independent Kenya'. *Comparative Studies of South Asia, Africa and the Middle East*, 26(3): 491–505.

Ong, A. 1999. *Flexible Citizenship: The Cultural Logics of Transnationality*. Durham: Duke University Press.

Open Society Justice Initiative. 2011. 'Nationality and discrimination: the case of Kenyan Nubians'. [Online] Available at: www.opensocietyfoundations.org/publica tions/nationality-and-discrimination-case-kenyan-nubians [Accessed 4 September 2017].

O'Reilly, K. 2000. *The British on The Costa Del Sol*. New York: Routledge.

Rodriguez, V., Fernandez-Mayoralas, G. & Rojo, F. 1998. 'European Retirees on the Costa del Sol: a cross-national comparison'. *International Journal of Popular Geography*, 4: 183–200.

Rogaly, B. & Taylor, B. 2010. '"They called them Communists then … what d'you call 'em now? … Insurgents?" Narratives of British military expatriates in the context of the new imperialism'. *Journal of Ethnic and Migration Studies*, 36(6): 1335–1351.

Stasiulis, D.K. and Yuval-Davis, N. 1995. *Unsettling Settler Societies: Articulations of Gender, Race, Ethnicity and Class*. London; Thousand Oaks, California: Sage.

Steyn, M. 2012. 'The ignorance contract: recollections of apartheid childhoods and the construction of epistemologies of ignorance'. *Identities*, 19(1): 8–25.

Stoler, A.L. 2016. *Duress: Imperial Durabilities in Our Times*. Durham: Duke University Press Books.

Torres, D.S. 2006. 'Elderly immigrants in Sweden: "Otherness" under construction'. *Journal of Ethnic and Migration Studies*, 32(8): 1341–1358.

Tsing, A.L. 2015. *The Mushroom at the End of the World: On the Possibility of Life in Capitalist Ruins*. Princeton: Princeton University Press.

Turner, V.W. 1967. *The Forest of Symbols: Aspects of Ndembu Ritual*. Ithaca and London: Cornell University Press.

Uusihakala, K. 1999. 'From impulsive adventure to postcolonial commitment: making white identity in contemporary Kenya'. *European Journal of Cultural Studies*, 2(1): 27–45.

van Bochove, M. & Engbersen, G. 2013. 'Beyond cosmopolitanism and expat bubbles: challenging dominant representations of knowledge workers and trailing spouses'. *Population, Space and Place*, 21(4): 295–309.

van Tol, D. 2015. 'The women of Kenya speak: imperial activism and settler society, c.1930'. *Journal of British Studies*, 54(2): 433–456.

Walsh, K. 2010. 'Negotiating migrant status in the emerging global city: Britons in Dubai'. *Encounters*, 2: 235–255.

Walsh, K. 2012. 'Emotion and migration: British transnationals in Dubai'. *Environment and Planning D: Society and Space*, 30(1): 43–59.

Walsh, K. 2016. 'Expatriate belongings: traces of lives "Abroad" in the home-making of English returnees in later life'. In *Transnational Migration and Home in Older Age*, edited by K. Walsh & L. Näre, 139–153. New York: Routledge.

Walsh, K. & Näre, L. eds. 2016. *Transnational Migration and Home in Older Age*. New York: Routledge.

Wang, L.K. 2016. 'The benefits of in-betweenness: return migration of second-generation Chinese American professionals to China'. *Journal of Ethnic and Migration Studies*, 42(12): 1941–1958.

Wasserman, G. 1974. 'European settlers and Kenya colony thoughts on a conflicted affair'. *African Studies Review*, 17(2): 425–434. https://doi.org/10.2307/523642.

Zontini, E. 2015. 'Growing old in a transnational social field: belonging, mobility and identity among Italian migrants'. *Ethnic and Racial Studies*, 38(2): 326–341.

6 Transgressing transnational normativity?

British migration and interracial marriage in South Africa

Daniel Conway and Pauline Leonard

Introduction

In this chapter, we reengage with the research we conducted from 2010–2013 with white British' immigrants in South Africa who were born in Britain and migrated to South Africa during the apartheid/post-apartheid period spanning the 1940s to 2000s (Conway and Leonard 2014). We draw on this ethnography to expand the horizon of research in British out-migration, which to date has tended to be dominated by a focus on white British migrants and white couples/families, particularly when considering lifestyle or privileged migration, (e.g. O'Reilly 2000; Benson 2011; Leonard 2010; Walsh 2007). While this has led to productive and critical analyses of the constructions of whiteness and British nationality in migratory contexts, there remains a gap in our knowledge of British migrants of diverse racial and ethnic backgrounds (although see Higgins, this volume, for an exception). However, in our research of the British in South Africa, several of our respondents provided a counterpoint to this myopia, through their marriage with black and Coloured[1] South Africans. The different national histories and racial politics of Britain and South Africa, underpinned by their relationship as coloniser and colonised, infuse these partnerships in terms of attitudes, experiences and lifestyles as well as beliefs about identity and belonging (Edwards 2017). Interracial marriage[2] can push differences in background to the fore, forcing British migrants to address what are often, for them, new questions of social and political affiliation and lifestyle. Further, mixed marriage poses a challenge to what we term the racialised 'transnational normativity' of (white) British migration to South Africa.

To further our analysis, we introduce the concept of 'transnational normativity' to conceptualise the ways in which, historically, white British migration to South Africa has been primarily fuelled by desires for, and promises of, upward social and economic mobility. In other words, as we go on to demonstrate, in the apartheid era British migrants were white, often attracted to South Africa by National Party government or multinational corporation policies offering subsidised transport, well paid jobs, high quality accommodation, good schools and enhanced social status (Conway and Leonard

2014). In the post-apartheid era, many of our research participants continue to enjoy life in South Africa because of the material and social benefits they believe it gives them: large and comfortable houses, cheap and luxurious food, domestic labour, relatively affordable private education and healthcare and beautiful scenery. Through and within this, a class-based British 'way of life' can be easily established in the new national context, with transnational performances and links maintained. While we found this to be 'normative' for the majority of respondents within our original research, we did also meet people who transgressed this pattern, turning away from the apparatus and associated privileges pertaining to the British migrant community. At the same time, however, as we discuss in this chapter, our analysis reveals that some of the benefits which accrue to whiteness continue to inflect the 'transgressive' British migrants' racial beliefs and practices, albeit in less overt ways.

This chapter starts by revisiting the issues surrounding the somewhat contested category of 'the British in South Africa' and the problematics of conducting research with this group. It then introduces the concepts of interracial partnership/marriage and 'racial projects' (Omi and Winant 1994) to develop the theoretical lens used to explore the narratives of Andrew and Caroline, two of our respondents who 'partnered out' (Caballero et al. 2008: 50). These are presented in some depth, to enable the full richness of their stories to be revealed. We end by concluding that, while the privileges which accrue to whiteness have clearly continued to frame their experiences as well as their accounts, their stories reflect important diversities within the category of whiteness and challenge the normative expectations of lifestyle enhancement held by many of their compatriots.

Researching the British in South Africa

Researching the British in South Africa directly engages with a broader social politics about whiteness, privilege, nation building and the historiography and contemporary politics of South Africa. However, classifying white British migrants in South Africa is difficult. The country's history of colonial settlement, the development of an 'English-speaking' white identity, related to Britishness, but not necessarily predicated on being descended from a UK citizen, makes the boundaries between 'British-born' and 'English-speaking' white South African a fluid one. Furthermore, the community has been joined at various times by 'Anglophiles' from outside the group, such as General Smuts and even Nelson Mandela, who have popularly been acclaimed for embodying the 'values' of British-liberalism and democracy (Dubow 2009). For some, the importance of discussing English-speaking whites lies in emphasising the community's presumed liberalism, its contribution to the anti-apartheid movement and its assumed benign impact on the country. Of course, such assumptions obscure the realities of English-speaking complicity and agency in enforcing racial exclusion and apartheid. For British born white immigrants, their presence in the country was an important facet of white minority apartheid governance; in post-apartheid South Africa British migrants have experienced the benefits and challenges of being white alongside other white South Africans. Alternatively, when we were conducting our research, we were also aware that for many, the British in South Africa are considered as simply part of a homogenously racist and declining group of white South Africans, hardly worth researching at all. In contemporary South Africa, anyone who researches and writes about whites and whiteness, and who problematises the widely held contentions about 'English speaking' whites: that they were not responsible for apartheid's racism, or that they have faded benignly into the background, can quickly come to be viewed as provocative (Conway 2016). Thus while our research on the British in South Africa conformed to many of the conventions of privileged and 'lifestyle migration', it also transgressed them, and pointed to the difficulty of applying such terms to all British communities across the world.

Ostensibly, many of the people we spoke to could easily be considered 'lifestyle migrants', migrating to the country for its sunshine, beautiful landscapes and opportunities for luxurious living at much-reduced cost. Yet although we were told narratives that expressed all these factors, these accounts often belied more difficult realities in everyday life: encounters with the country's high crime rate, fears for the nation's political future, the cost and necessity of health insurance, weak South African exchange rates and property price differences that severely constrained prospects for return. Even the imperialistic enjoyment and appropriation of the country's landscape and natural beauty was often simultaneously accompanied by a reluctance to walk in the landscape and an acknowledgement (or experience) that it could be dangerous. This raised questions about how 'privilege' was constructed in racial, social and economic terms. The British in South Africa benefit and help perpetuate considerable racial privilege, enjoy varying degrees of social privilege, but economic privilege and the ability to be globally mobile varied considerably (Andruki 2010).

When we commenced our research, we were initially swamped with offers for interview participation and it quickly became clear that the people we spoke to did indeed identify as British in ways differential to just 'white', 'English speaking' or 'South African', and that as a group they were difficult to characterise (Conway and Leonard 2014). There was a significant difference between areas of residence in the country – many 'lifestyle' migrants had chosen the tourist landscapes of the Western Cape, whereas many professional/labour migrants had chosen the 'real Africa' of urban Johannesburg. In between were a mix of ex-colonial residents who had migrated south as the British empire retreated from Kenya, Zambia and Rhodesia [now Zimbabwe], working class 'assisted passage' migrants from the 1960s to the 1980s, post-apartheid retirees and young professionals or entrepreneurs who had visited the country first as tourists and had returned to establish a new life in a dynamic new democracy.

In our book, we explored the lives of this wide range of British migrants. We documented many instances of racism, partialised and uninformed accounts of the country's apartheid past, and lives lived in white enclaves, quite divorced from the society around them. For some, the ability to sustain this racially exclusive and privileged lifestyle was openly acknowledged, celebrated and a contingent condition for remaining in South Africa. Yet there were some who did not conform to what could be considered as the 'transnational normativity' of white migrants in South Africa and privileged migrants more generally. It is these migrants who we consider in this chapter. In particular, we will consider the extent to which their whiteness diverged, destabilised and reconfigured the norms of whiteness in South Africa. To what extent was their privilege disavowed or reshaped? What can their lives tell us about the gaps and oversimplifications of the conceptualisation of the privileged white migrant? Contrastingly, we will also analyse their self-narratives and everyday lives against the claim that even privileged migrants and white residents of South Africa who do adopt different political and social positions and actively position themselves against other whites, merely obscure their ongoing privilege and are disingenuous.

Whiteness, privilege and interracial marriage

White British migrants in South Africa live in spaces powerfully shaped by the apartheid past and the legacy of discourses of whiteness continue to sustain the contemporary context. These discourses, emanating from the South African media, universities, politicians and everyday life, shape 'how the world comes to be known' and thus their exploration 'opens up useful ways of exploring the intersection of race, space and identities' (McEwen and Steyn 2013: 2). In our research, it was clear these discourses of whiteness shaped senses of belonging or un-belonging, where people chose to live, work, walk, drive and shop. Contemporary scholarship on white South Africans and whiteness as a set of discourses broadly concurs that whites have failed to accept responsibility for their complicity in apartheid or to genuinely engage with South Africa's nation-building project. For example, some members of the white community openly rejected, ignored or derided the Truth and Reconciliation Commission[3] and have subsequently chosen to misrepresent or deny knowledge of its findings (Conway 2016; Steyn 2012).

Crucially, although conservative and openly racist discourses of whiteness do continue to exist in South Africa, they have been predominantly replaced with white liberal discourses that ostensibly embrace the values and project of the new multi-racial South Africa, yet which simultaneously obscure, reject the reality of, and thus perpetuate, ongoing white socio-economic privilege (Conway 2016; Steyn 2001; Steyn 2012; Steyn and Foster 2008). Such 'white talk' (Steyn and Foster 2008) repudiates individual complicity in racism and can even argue that the racism of apartheid was not 'as bad' as claimed (Steyn 2012; Leonard forthcoming). In our research, we both spoke to British migrants who sought to deny that they had ever been racist, that the BBC had exaggerated the realities of apartheid and that even claimed that the country had not experienced a State of Emergency during the 1980s and 1990s. Such claims 'are premised on intentional ignorance about the past and also a desire to ignore and discount inconvenient and disruptive perspectives, arguments and facts' (Conway 2016: 120). Mills (1997) argues this ignorance is integral to the racial contract that perpetuates white privilege: 'One has an agreement to *mis*represent the world. One has to learn to see the world wrongly, but with the assurance that this set of mistaken perceptions will be validated by white epistemic authority' (p. 18). Listening to accounts that denied, obscured and ignored the realities of apartheid and/or the racial and socio-economic inequality of contemporary South Africa were remarkable, but unsurprising given the political and racial stakes involved – British migrants who perpetuated such discourses clearly conformed with, and were complicit in, dominant discourses of whiteness in South Africa that seek to maintain and perpetuate white privilege.

Yet what of British migrants who do not reproduce 'white talk' (Steyn 2001) and whose everyday lives seek to repudiate and reformulate whiteness? It is these migrants we wish to discuss here. We both spoke to British migrants who actively sought to step outside the discursive and physical constraints of whiteness, whether by living and working in multi-racial, or black, spaces and jobs, rejecting a 'privileged' lifestyle based on the subordination and servitude of black South Africans, or engaging in equality-based politics and relations. Here our interest focuses on two respondents who had 'partnered out' (Caballero et al. 2008: 50) with black and Coloured South Africans. In South Africa, interracial relationships, both past and present, have reflected the social politics of race by subjecting participants to greater levels of scrutiny and judgement (Du Toit and Quayle 2011). Thus, while interracial marriage is by no means an indicator of a diminishment of social distance between ethnically bounded groups such as whites and blacks, research in South Africa has found that interracial marriage and families are a still rare, but significant and sustained form of cross-racial contact that reduces prejudice (Du Toit and Quayle 2011). In terms of critical studies of whiteness, interracial marriage can provide a significant lens to the internal diversities and changes in outlooks, politics and racial and ethnic awareness within the category of 'white' (Song 2016).

South Africa has a long history of interracial unions since the Dutch settlement of the Cape in the seventeenth century. For the colonisers, there was a shortage of women and, despite policies to ship out British women of marriageable age (Van-Helten and Williams 1983; Leonard 2013) white men commonly married black women up to the start of the twentieth century (Jacobson et al. 2004). The turn to racial segregation brought approbation to such marriage and, as such, the outlawing of interracial sexual relations became a key tool of the apartheid regime. This was achieved through the Mixed Marriages Act (1949) and the Immorality Act (1950). Daniel Malan, the South African Prime Minister who championed Afrikaner nationalism in the 1948 elections, saw it as a fundamental mechanism by which to sustain white power and segregation (Jaynes 2007). The Mixed Marriages Act was repealed in 1985, but interracial marriage is still relatively rare in South Africa today: the overall odds of a person marrying someone of the same race group have dropped from 303:1 in 1996 to 95:1 in 2011 (Amoateng 2015). However, while Amoateng suggests that about 5 per cent of Coloureds, Asians and Indians marry outside their ethnic group, whites remain the least likely to do so. This is compared to 9 per cent (and rising) of people in Britain living as part of a multi-ethnic relationship, with black/white marriage the most common (Song 2016). In some contrast to South Africa, 'the British state has always taken a more laissez-faire, albeit deploring, approach to relationships across racialised boundaries' (Edwards 2017: 179). Despite racial prejudice and some social condemnation, anti-miscegenation legislation was never a feature. Indeed, in postcolonial Britain, it has been argued that 'cultures of mixing' (Callbero et al. 2008): interracial and interethnic unions and the children borne of these, increasingly comprise normal, everyday life in many urban and suburban regions (Gilroy 1993; Song 2016). In contrast, South African dismantling of neighbourhood, residential and other forms of segregation has been slow (McEwan 2013; Du Toit and Quayle 2011).

These different social, historical and political contexts for the construction and transformation of racial categories and everyday racial meanings are reflected in Omi and Winant's notion of 'racial projects' (1994; see also Edwards 2017). A key aspect of their theory of racial formation, which addresses the complex relationship between historically situated social, economic and political forces in the construction of racialised hierarchies, categories meanings and identities, racial projects are practices and processes which occur at national, institutional and individual levels. The notion embraces the negotiations of 'race' in everyday life, the ways in which people position themselves within racialised discourses to produce meanings, practices and (racialised) identities. As noted above, these not only inflect where people chose to live, work, socialise and shop but also with whom they choose to have personal relationships and form families.

This framework provides a critical lens by which to explore the lives and experiences of respondents who 'partnered out'. For some of our respondents, 'partnering out' was experienced when marrying into white Afrikaans-speaking families: a complex negotiation of class, historical and linguistic difference, but still essentially a negotiation framed around whiteness and overall privilege. In contrast, we focus here in-depth on the narratives of two of our (white British) interviewees who were both partnered to black South Africans. As we now turn to explore, the intersections of nation and race, as well as gender and class, are evident in the ways in which they sought to transgress the transnational normativity of white British migrant lives.

The social politics of migrating to South Africa

Both Andrew and Caroline started their interracial relationships in the 1980s, when the sexual and racial politics of South Africa were beginning to shift with the governing National Party's so-called *verligte* [liberal] agenda of introducing limited constitutional reforms and removing some of the 'pettier' aspects of apartheid. However, in no way did the repeal of the Mixed Marriages Act in 1985 reflect a more general softening of apartheid from the National Party; rather it formed part of a series of tokens by which some appeasement of both the mounting international pressure and rising black militancy was attempted (Mann 1988: 54; Schrire 1991: 29–35). If anything, minor and often contradictory, efforts to reform apartheid only hardened and broadened opposition to white minority rule both within South Africa and internationally. In 1983, the newly established United Democratic Front (UDF) mobilised millions of black South Africans in protest against the government's proposals to grant limited voting rights to Indian and Coloured South Africans and none to black South Africans. These protests quickly spread across townships in South Africa and broadened to oppose every facet of white minority rule. The most sustained uprising against apartheid in the country's history had begun. The government responded quickly and decisively by sending the South African Defence Force into unruly townships, using the police to arrest or kill anti-apartheid activists and suspected members of the ANC and declaring successive States of Emergency. Other aspects of apartheid began to collapse from the mid-1980s onwards, the Group Areas Act 1950, a fundamental tenet of apartheid that restricted black South African access to urban areas and defined spaces as 'white' or 'black', began to be flouted. This affected Johannesburg, in particular, where a lack of housing and the weakening grip of the government led to thousands of black South Africans moving into former 'whites only' areas, particularly in the inner city. Neighbourhoods such as Hillbrow, that had high density whites-only housing and were particularly popular with young white immigrants, saw black South Africans moving in and the district developed a noticeable sub-culture, with interracial bars, friendships, gay bars and shared housing becoming commonplace (Conway 2009). These shifts were by no means smooth or welcomed by the majority of whites, many of whom 'fled' the inner city and moved to Johannesburg's 'whiter' northern suburbs, or emigrated (Conway 2009). The National Party denounced these trends and sought to re-impose the Group Areas Act but, as with their other efforts to maintain apartheid and suppress protest, sustained pressure from domestic protest, international sanctions and internal divisions made white minority rule untenable. Nelson Mandela was released in 1990 and negotiations began to transition to non-racial democracy concluded with the 1994 elections.

Simultaneously, in Britain, the 1970s and 1980s was also a time of growing popular resistance to apartheid South Africa. The incidents of the 1960s such as the Sharpeville massacre, where 69 black demonstrators were shot dead by the South African police, and later, the 1976 Soweto Uprising and particularly the repressive chaos of the 1980s had sent shock waves across the more liberal sections of the British public and support for the Anti-Apartheid Movement grew steadily over the ensuing decades as the media reported on and broadcast scenes of the black struggle (Conway and Leonard 2014). As boycotts of South African goods, services and sporting links intensified, it would have been difficult, for British audiences, to escape knowledge of the apartheid system and the structural inequalities it imposed (Conway 2018; Conway and Leonard 2014). At the same time, however, demand for new white (British) migrants to shore up the South African economy burgeoned during the 1960s and 1970s, and the National Party government supported an assisted passage scheme aimed to facilitate migration at little cost. Engineers, dentists, architects and public-sector workers were all in demand, and the nation's attractions of beaches, *braais* [barbecues], high wages, sport and good education were wielded as levers by which to attract white British migration (Conway and Leonard 2014). Andrew is one such engineer who migrated to the Free State from London in the early 1980s, and it is to his story we turn to first. This is followed by Caroline, who met her South African partner in the UK, in the progressive and anti-Thatcherite social political space that was Manchester in the 1980s. Caroline joined her husband and one of their children to migrate to in the 1990s, following the end of apartheid.

In both their narratives, far from being stable and/or a strategic set of ideas and attitudes, their individual racial projects are revealed as highly contingent upon the broader national racial projects in which they are situated. Consequently, just as these social and political contexts change quite dramatically over the timescapes of their accounts, so too are their racial projects revealed to shift over time. In both cases, the comparative national racial contexts: the ever-increasing racial integration of multicultural Britain and the intensified systems and practices of racial segregation which exist in South Africa resonate in Andrew and Caroline's talk. As they both become more aware of the daily realities of the South African context to which they had migrated, their racial projects become increasingly politicised, transgressing the separatist transnational normativity of many of their compatriots.

Shifting whiteness and rejecting lifestyles

Although the easy availability of work in the construction industry was the channel by which he came to South Africa, Andrew admits that, from the start, his migration was always more of a racial project:

> To be perfectly honest, I came here really wanting to see what apartheid was like. Which upset a few people in my family because they didn't really want me to have anything to do with South Africa.

Andrew acknowledges that his knowledge of apartheid was sketchy in the early days of his arrival, as indeed it was for the other engineers he was rooming with:

> All three of us knew that the black people maybe didn't live in quite the same style as the white people. But we really didn't have any idea whatsoever. One of the guys was from Australia and his idea was that the black people lived very well, they had their nice houses. And I still argue with him today, I say, "You obviously didn't have a clue, did you?"

They were soon starkly awakened to the racially segregated realities of daily life. Their work took them into the middle of the gold mining area where:

> there was absolutely no mixing between whites and blacks and it's very, very segregated even the shops. Not just the government institutions like post offices and government buildings and things like that which were segregated throughout the country. Even small shops, they'd have like the normal sort of white area and then a little sort of window with a ledge where the black people could go and buy their loaves of bread and things like that. And that was a real eye opener.

Andrew's work building a school took him into the black township of Soweto. He was fascinated and from the start wanted to forge friendships with his black co-workers. Coming from Britain, the various technologies of governance of the apartheid regime struck him as he nonetheless defied them:

> So I used to go down to the black quantity surveyor and the engineer and whatever and whoever, I could go round their houses because I had the permit to go into the township. Otherwise you didn't go into the township and you would get into trouble if you did. I don't know what it was like in Jo'burg. I think Soweto was – they had the signs up, I can remember this big green sign saying, 'permits'! I've got a photo some-where. It was sort of in your face but Soweto was so vast that you could actually go in, even in 1981. In a Free State town that was fur-ther out, you would just get in a lot of trouble with the police and certainly thrown out, that's for sure.

Despite the legislation forbidding white engagement with places such as Soweto, Andrew was determined to find out what was happening, politically and socially. He enjoyed visiting for both work and leisure 'because it allowed me to see what I was really going to see. So I was very lucky in that respect'. His friendships in Soweto strengthened, although developing these under the apartheid regime was a constant challenge:

if we wanted to go away and socialise we'd go into Lesotho which was very close, in Swaziland, because there, there was no segregation. So we ... just as you would with any group of friends ... you get in a minibus or a couple of cars or whatever and off you go. And I used to travel with them in the same vehicle, which might sound like it's nothing – but there were laws.

Andrew is, as he puts it, 'not shy of drinking' and his visits to Soweto took him into the various *shebeens* (bars) where he thoroughly enjoyed the atmosphere: 'it was great, because you really felt what African South Africa was like'. He recognises how unusual his approach was at the time, surmising that his white colleagues 'had no concept of what was going on then, or now for that matter in the townships, they just had no idea'. In contrast, Andrew's motivation was *not* to live in 'the nice white houses with their swimming pools, then half a mile away the other side of the electricity pylons was the township which was all, in those days it was dusty roads, no cars, and little boxes that they used to live in'. Thus, while Andrew's work meant that he was housed on site in rudimentary accommodation with the other white migrants, he was not unhappy with this. *His* racial project was to live a culturally mixed life, regardless of the fact that he was the only man in his white circle to do so and of the potential consequences:

> I didn't think anything of it in those days. Didn't have any fear of getting into trouble or being caught doing something that I could be told off for. Didn't really cross my mind because it was like, well that's what I wanted to find out about, and I did.

Despite Andrew's declarations of difference from his compatriots, Andrew's own privilege remains unacknowledged or recognised here. In that it did not occur to Andrew that he should not have access to any space he wished to enter, or that a white police force might be hostile towards him, Andrew shares the same sense of entitlement performed by whites from across the political spectrum.

Andrew's disregard of any boundaries to his rights to investigation led him to the heart of Soweto's *shebeens*, where he met his wife-to-be, who ran one such business. While their developing relationship contravened legislation and white normative attitudes, the changing political context and the dynamism for a new way of life that was emerging within Johannesburg at the time is reflected in Andrew's memories:

> Not at that time that this really was an issue for me, but as a white man, a black woman was not allowed to ride in the front seat of a car, she had to go in the back seat. So I mean, there were ways you could get caught, and actually later on I did, you know, driving around with my wife to be and we'd get stopped, and it would be, "What's she doing in there?" We're

just going to wherever it might be, off on a journey somewhere. By that time, this was like '85/'86, it was getting easier, particularly around Jo'burg, it was actually becoming quite cosmopolitan. So there wasn't really much they could do. But they still they were pretty nasty. Gave you a hard time I'd guess you'd say.

It is possible Andrew over-exaggerates the extent of change in attitudes beyond Hillbrow and the more liberal Johannesburg suburbs. For, less than six months after arriving in South Africa, Andrew moved out to Soweto to live with his girlfriend and her child, rather than she move into Jo'burg. It was here that Andrew made a new discovery, after seeing a name he recognised on a letter addressed to her son. It was revealed to him that his partner was in fact a family member of one of the most senior and renowned leaders of the United Democratic Front (UDF). As Andrew explains, the UDF had, in part, been established as a cover for the ANC, which the National Party government had banned. His partner's high-level political connections and the intensifying political context meant that their life was an unstable existence, with Andrew needing to be prepared to flee at a moment's notice for the safety of his new family:

> I was rumoured to be one of perhaps two or three white people who were in one way or another living in Soweto. Because actually you couldn't really live there, if things got tense, which they did from time to time in, that was really 1985 when I spent most of my time there, then you'd just have to leave because it was too dangerous. Not so much for me, but for my wife-to-be's family. You didn't want to attract focus, or attention.

Although this was far from the racial project with which Andrew had initially set out: the more socially driven ambition to live a culturally mixed life, he now shifted towards active political participation, Andrew's relationship meant that he became centrally involved with the missions and activities of the UDF. Rather unexpectedly therefore, his life took a new turn, prompting him occasionally to think:

> 'What am I doing, getting involved in this kind of thing?It wasn't really quite what I intended! So it was really bizarre the way it all happened. No deliberation about it really ...

As Andrew's story continues, he reflects the fact that, while they were few and far between, white people were centrally involved in the struggle:

> And then of course it was one thing after another. Winnie [Mandela] got released from house arrest in Brandfort in the Free State. And we started to get to know her and then we started to get to know ... God I can't remember now, various ANC type people. But one of the people we used

to go and see a lot was Helen Joseph who was quite a famous activist. She was actually British, but she'd lived here since the '40s or whatever. She was under house arrest in Park Town, Norwood [Johannesburg].

The intensive political scrutiny placed upon covert ANC and UDF activists as the anti-apartheid movement gained momentum became increasingly threatening as the 1980s progressed. White activists were not immune from these threats. Indeed, the death in police custody of the white trade unionist Dr Neil Aggett in 1982, alongside the assassination of the Ruth First, a South African academic in exile in Mozambique, sent clear signals of the government's repressive intent. By the mid-1980s, white anti-apartheid activists, such as those in the End Conscription Campaign, were subject to regular police surveillance, harassment and detention without trial (Conway 2012). As well as threat of being caught up in, or even the target, of police or army violence in the township, Andrew would have faced possible imprisonment and deportation had he been investigated.

The constant risks to the security of Andrew's partner and child meant that they felt it safer to move to Lesotho, where they could get married and all live together more freely. Nevertheless, they still 'came backwards and forwards up to Jo'burg maybe once a month and we'd go and meet Winnie and whoever … it was all a bit surreal really'. As Andrew's relationship solidified and he became step-father to his partner's child, his racial project, once again, takes a further shift, to include being a husband and father in an interracial family: 'To me it was all just like that was our family, I wasn't really taking them as being supposedly important people. And by that time my step-son, he was just an ordinary kid …'. With this statement however, Andrew demonstrates that he simultaneously attempted to transcend race, nationality and racial identification within his family relationships, focusing instead on just being a 'normal' family.

Yet at the same time the political context framed their lives. Fighting apartheid was a key part of their identities and philosophy, so Andrew said to his wife:

What do we want to do really? Do we want to just be able to live in Lesotho or do we want to live in Jo'burg? Which is really where we'd prefer to live. So, we had a look around Jo'burg to see whether it was feasible, knowing that obviously it would be difficult even in '87. Basically, it wasn't really practical. The only place that we could have lived in those days was in Hillbrow, which was sort of a party place, and it was the first to sort of mix. So by '87, instead of being whites-only which it was in '81/'82 it had started mixing. And obviously the first thing that crossed my mind was, there's a lot of black people around here and they're quite free to move around, whereas previously there very much hadn't been. So that was the only place that we could find anywhere to stay. And even that, it wasn't legal. What do they call it, the Group Areas Act was still fully enforced.

As previously discussed, although the Mixed Marriage Act was repealed and the Immorality Act had been amended in 1985, the Group Areas Act meant that mixed couples were still not able to live with each other legally: just one of the contradictions that enabled apartheid governance to be maintained. The family's struggle to live a 'normal' life finally prompted them to move to Britain at the end of the 1980s, where Andrew's wife 'settled down really well, she slotted in straight away, she's very gregarious and she got involved with everybody and never had a problem in England at all'.

Although happy in England, the marriage eventually ended civilly. The two are still on good terms, with Andrew still acting in a parental capacity to his ex-wife's son. Andrew has since married again to a South East Asian woman with whom he has had two children. As a consequence, it was some time before he returned to South Africa, doing so after being offered a new job in 2008. Arriving back in the country, he was struck by the changes in lifestyle and social relations:

> just amazed, you see black people and white people and they're not always at different tables, they're very often at the same table, and believe me you never saw this before. And even these days as well you see school kids, black school kids who speak with a white South Africa accent. It amazes me every time, even now. I still double take when I hear it. Fantastic.

These perceptions, while genuine, are of course predicated on a considerable level of privilege: the ability to dine at restaurants and also attend previously whites only schools. While Andrew applauds how the political and social context has progressed, he is less convinced that his mixed-race children will thrive in South Africa. Belying his post-race rhetoric, he explains that they have experienced some racist taunts at school. When I last spoke to him, Andrew reveals that he has engaged in a new racial project: contemplating moving to his wife's home country, so that his children might enjoy a better sense of national belonging and identity. Once again however, Andrew's racial projects are grounded in his ability to move unfettered across the globe.

Spatial, racial and political journeys

'We make psychological journeys through making physical journeys', reflects Caroline, whose life as part of a mixed-race family and occupation as a development worker with the poorest South Africans is quite different to many of the other white British we interviewed. Caroline's identity, like Andrew's, was also framed and invested in deep political commitment, albeit she had expressly not emigrated to South Africa in the 1980s because of this. Rather, she moved to Manchester to escape the conservative confines of her background in the Home Counties and participate in the city's progressive opposition to the right-wing Thatcher government. As a social worker in

Manchester, she recalled working 'when issues of race, sexuality, disability and environment were actively being addressed by a sufficient number of people to make a difference'. In Cape Town, she continues to work with the economically and socially marginalised and this political and social engagement remains very important to her. As such, Caroline's development work can be considered a type of 'moral labour' (Fechter 2016) that traverses and defines both her personal and professional identities.

Caroline began a relationship with a Coloured South African man she met in Manchester in the early 1980s and they adopted children of Jamaican heritage, yet she had not planned to ever move to South Africa. Indeed, her first visit to the country in the late 1980s was an 'incredibly uncomfortable experience' because the mixed-race family 'felt too stared at … incredibly scrutinised'. Caroline's husband travelled to South Africa to witness the country's first democratic and non-racial elections in 1994 and she instinctively knew he would return and say, 'he wanted to go home'. 'My family is complex', explained Caroline, because of the Coloured, Jamaican and now black African mix (her grandson is mixed Jamaican/black African) and, because of this, Caroline's racial and national identity is complex:

> I'm very clear I'm white and I'm very clear that being white in South Africa is very different to being white in England … I live in a Coloured extended family. It's not black, but that's not a political definition of black, and for me it is [black]. I've lived in a black family longer than I haven't.

Caroline's family, her commitment to working with the marginalised communities in Cape Town and her choice to live in a mixed-race area of the city, rather than one of its well-heeled predominantly white areas, is part of the reason she feels disassociated from other white British migrants in South Africa: 'I don't view myself as part of a wider body of British people particularly. I don't have a national identity framework going on'.

Caroline recalled that before moving to Cape Town in 1998 in her early 40s, they sold their house in Manchester and

> it felt like a severing of something … and it was incredibly risky. We didn't have anywhere to live, I didn't have the right to work, we had to get her [Caroline's 11-year-old daughter] into school, we were staying with his [Caroline's husband's] relatives.

With the assisted passage scheme having ended in 1991 and the incentive to hire white immigrants gone, Caroline and her family lived in financially straightened circumstances for quite some time. Eventually, Caroline joined the *Big Issue* homeless project in Cape Town, drawing from her experience and skills as a social worker and helping the British charity adapt their policies to the local needs of Cape Town's homeless population. Yet it was not an easy transition, Caroline's daughter was homesick and wanted to return and, as she explained, adapting to life living with an extended Coloured family in Cape Town was challenging. Indeed, she recalled 'I didn't feel at home in that particular community at all and I was outraged by some of the stuff I heard and', as shall be discussed below, 'some of the things that happened. Really offended!'.

Caroline was aware that her race and nationality influenced her husband's welcoming and positive attitude to her. To them, he was 'marrying up, which is marrying white'. Caroline found herself part of a very regimented set of daily routines that often revolved around class (and racial) prejudice that reflected the spatially classed and racial topography of Cape Town: 'it's only nice if you have friends in Pinelands [a predominantly white upper class area] and you must go shopping in Rosemead Avenue [a predominantly white upper class area] etc. etc.' she recalls her husband's family repeatedly telling her. 'It was about race, it was about class, it was about all that sort of very aspirational essentially reactionary kind of lifestyle. I was outraged by some of the stuff I heard and some of the things that happened', remembers Caroline. For example, one trip to the shopping mall with her partner's Aunt angered Caroline considerably:

> she [the Aunt] used to try and shoot through the traffic lights to avoid beggars 'and on this occasion she was not quick enough. A beggar approached the window and asked for money, She said 'no I can't give money'. She said I was her Madam [colloquial term for white female employer of domestic workers] and she was driving my car and I was like you can't do this kind of thing and I was *so* angry ... and I think with my kind of English honed anti-discriminatory approach ... I was shocked to the core and ... I didn't like being drawn into it, being made complicit.

In political and social terms, life could hardly have been more of a contrast to her previous existence in Manchester, but in many respects her experiences with her husband's family mirrored the broader politics of the post-apartheid Coloured community in Cape Town. Coloured voters have joined whites in supporting the opposition and broadly conservative Democratic Alliance party, who now govern the city and province. For many of the 'lifestyle' migrants we interviewed in Cape Town, this was a main reason for living in the area. Indeed, some happily described the Western Cape as 'not the real Africa' and Cape Town as, even, 'England by the Sea' (Conway and Leonard 2014). The social and political 'white space' of the Western Cape was accompanied with a sense of a luxurious Mediterranean lifestyle, albeit at a more affordable price than could be obtained in the south of France or Italy. However, Caroline's 'England' was somewhat different to the provincial and conservative 'England' of some of our other informants and she was determined not to live isolated from the multi-racial society surrounding her. As she explained, 'if you're somebody who chooses to make connections and that's what I chose to do in Manchester and that's what I choose to do here as well'. Working for a UK-based homeless charity project 'I could hardly have had a better introduction to some of the poorest South Africans' and after moving out from her husband's family's home she began to settle.

As discussed above, most of the British migrants we spoke to, particularly in the tourist landscapes of the Western Cape, premised their sense of enjoyment and belonging in South Africa on the enhanced material possessions the country gave them. This is a key part of what, after all, defines the 'transnational normativity of their' expectations and experiences, and forms the basis for their definition of 'lifestyle migration' that delivers an enhanced and enjoyable standard of living for British migrants (Benson and O'Reilly 2009). Even for those who consciously engaged with, and were troubled by, the country's economic inequality and racial exclusion, tended to do so firmly from white enclaves. For Andrew, living in Johannesburg, reaching out to black South Africans and witnessing the forefront of the struggle against apartheid was an opportunity he could and did choose to take. Today, many of the respondents we spoke to continue to contrast multi-racial, metropolitan Johannesburg as being the 'real South Africa', whereas Cape Town was perceived to be a colonial, parochial and quintessentially 'English' outpost. As explained above, this epithet about Cape Town was exactly as some of our respondents in Cape Town wanted it, but Caroline also identified this contrast between Cape Town and Johannesburg and believed she had subverted it:

> I think it's still the case that unless you choose to live in one of the fusion areas, Cape Town is still a lot more segregated than Johannesburg by race and it's quite difficult to break through those boundaries. I feel as though I have managed to do it, but it's still not as easy as I'd like it to be.

Today Caroline lives in the Observatory district of Cape Town, which is a mixed-race and in Caroline's terms a 'fusion area', and for her this is 'what it means to live in a way that is inclusive, as opposed to living in some kind of white ghetto'. Residing in a 'fusion area' was also matched by using space differently. For most of our respondents in the Western Cape, landscapes were admired, indeed they were often cited as the primary reason for their sense of belonging in South Africa, but these same landscapes were also feared and seldom ventured into. For Caroline, using public transport, working and travelling through some of Cape Town's poorest communities was an everyday reality. 'I have a car', said Caroline, 'but I still get on [communal minibus] taxis. I am usually the only white person and the only white woman in the taxi'. In school holidays, Caroline had taken her grandson to the beach in predominately Coloured areas and as she recalls her partner said, 'you realise you're one of the only white people on the beach' and I said 'yeah, I'm not uncomfortable with that'. She remembered that in the past the realisation of this racial 'standing out' *had* been uncomfortable for her daughter when a teenager and she wondered if the same could happen for her grandson, but it was a phase she believed would pass.

As in Manchester, Caroline's experiences with working with the marginalised had influenced her politics in personal and societal terms. At the *Big Issue*, vendors would come to work and 'wouldn't have eaten'; they 'literally had nothing to eat'. This had a profound effect on her, 'I learned stuff that has changed me. I will never unthinkingly eat food or throw away food and I share food … I'm prepared to have less in order for other people to have more'. On return visits to the UK, Caroline became aware of British society's consumerism and materialism, something she found unethical and that she decided to reject:

> Something about me has shifted. I personally feel I need less, I actively want less. I know I can survive with much less. Lots of South Africans don't do that. There's lots of money in South Africa, but we don't have to do it. We had very little for a very long time. A friend came and visited and said 'Where's your cooker?' and I said 'we haven't got one' and she said 'Oh my God! I must buy you one!' and I said 'No, it's OK. We've got a camping gas stove. It's OK and I'm happy with that.' It's actually very liberating.

This rejection of materialism is a choice that poorer South Africans cannot choose: it is enforced on them. However, Caroline's choice is in stark contrast to the majority of British we spoke to in Cape Town, who's enjoyment of life was predicated on enhanced consumption and material comfort. Caroline could be considered a lifestyle migrant, albeit one that rejects the material comforts of whiteness: 'I came here for a different life', said Caroline, 'In fact I do have a better life. I have a simpler life and that is better'. When thinking about whether she missed the UK or whether she could imagine returning one day, Caroline reflected that the challenges, complexity and work she had to do in Cape Town was 'enough to keep me going for my lifetime' and she concluded 'We're complex ourselves. We fit in here very well. Probably more easily than we would there [in Manchester].'

Discussion and conclusion

British migrants who partnered across racial lines disrupted the norms of whiteness in South Africa and, perhaps as a consequence, continually and reflexively considered their racial and (trans)national identities. But to what extent was their privilege disavowed or reconfigured? Conceptualising these discursive, embodied and material acts as transgressive or transformational is the subject of some debate in critical race studies. Marriage and partnering across race, moving across racial spatial and social divides and proclaiming solidarity with racial 'others' is a far from radical act from the perspective of some critical race scholars. While space prevents us from a focus on *sexual* desire in this chapter, scholars have demonstrated how this is both gendered and racialised, part and parcel of how the colonial project was lived (McClintock 1995). We have focused on marriage and romantic partnering across race because as a public, ongoing and social act that has the potential for racial change. However, for Applebaum, 'racism is often perpetuated through *well intentioned* white people' (2008: 297). For example, when choosing to live in a neighbourhood that is not 'all-white … the very choice assumes and reinforces the "privileged choice" they have. Privilege is something white people tend to assert even as they seek to challenge it' (Applebaum 2008: 294). Sara Ahmed believes even open acknowledgement of past racial injustice by whites and a desire to reconcile, or express solidarity with black political struggles 'can work to *block* hearing' and displace whites and whiteness as an object of critique (Ahmed 2005). Indeed, discourses of white anti-racism can become 'a matter of generating a positive white identity that makes the white subject feel good. The declaration of such an identity sustains the narcissism of whiteness' (Ahmed 2012: 170) and is, arguably, 'part of the problem rather the solution to systemic racism' (Applebaum 2008: 17). Proclamations of anti-racism and acts of solidarity can do little to change whiteness as a mode of privilege and exploitation.

During apartheid and in post-apartheid South Africa there is a clear tendency for some whites, particularly white liberals, to proclaim non-racism and solidarity with the Liberation Movement of the past or South Africa's new non-racial democracy, while simultaneously defining the terms by which limited socio-economic reform can take place, denouncing threats to white privilege and stymying genuine socio-economic transformation and democratisation (Conway 2016; Steyn 2012; Steyn and Foster 2008). This 'New South Africa Speak' as Steyn and Foster (2008) term it, is a particularly politically and socially conservative tactic pervasive across South African media, opposition politics and academia. When considering the accounts of Andrew and Caroline, questions of whether their lives and identities conform to the strategies of whiteness discussed above, or whether they are indeed genuinely transgressing transnational normativity and challenging and shifting whiteness come to the fore. Both Andrew and Caroline's accounts reveal the agency of first being able to migrate to South Africa (albeit for Andrew, with greater material resources) and the choice of where to live, who to meet and what social activities to enjoy; choices that were and are not afforded to all South Africans, or indeed many other immigrants to South Africa. The desire to *know* about other South Africans and other racial groups can also be understood as part and parcel of the English colonial project in Africa and elsewhere. Certainly, all of our respondents claimed to both *know* about South Africa and South Africans, and most claimed to not be racist. Yet although both Andrew and Caroline's choices reveal the social and material privileges of whiteness, they were and are both political choices and choices that directly threatened those same social and material privileges. They also remain in complete contrast to the choices made by the majority of white British living in South Africa and the majority of white South Africans (Du Toit and Quayle 2011). By marrying across racial divides, Andrew and Caroline are reminded daily of the social-politics of race in South Africa and of their own whiteness. Further, whereas many white South Africans who openly allied with the Liberation Struggle in the 1980s now expect to be rewarded and venerated for their commitment and are often sharply critical of the new dispensation (Conway 2016; Steyn 2012; Steyn 2001; Steyn and Foster 2008), neither Andrew nor Caroline make such demands or claims. Rather, again in contradiction to many of the respondents we interviewed and the majority of white South Africans, who struggle and expect to remain at the centre, their whiteness has been decentred and they readily acknowledge their minority and immigrant status in the new South Africa.

Finally, our discussion of Andrew and Caroline's lives makes a contribution to nuancing and, to some extent, counterbalancing, some of the dominant themes within contemporary research on British migrants. This predominantly focuses on migrants living in privileged and often socially and racially segregated contexts, remaining distanced from political engagement and active citizenship in the host country. In contrast, Andrew and Caroline's interracial partnerships drew them into becoming active in national politics and embedded in working for social change and racial cohesion. Their stories are important not only when 'taking stock' of contemporary democratic South Africa (Steyn and Ballard 2013), but to furthering understandings of the complex and diverse dynamics of British race and nationhood-making in global contexts.

Notes

1 Mixed-race South Africans.
2 Following Song (2016: 632) we use the term 'intermarriage' to include a variety of unions between people of disparate racial, ethnic, religious and/or national backgrounds who are long term cohabiting but not necessarily married.
3 The Truth and Reconciliation Commission (TRC) was established by the Promotion of National Unity Act in 1995 to bear public witness, record and investigate evidence of crimes and violations of human rights in South Africa between 1960 and 1994. The TRC also had powers to grant amnesty to perpetrators of such abuses if they met certain conditions. The TRC reports, published in 1998, provided a comprehensive account of apartheid governance and abuses (Amstutz 2005).

References

Ahmed, S. (2005) 'The non-performativity of anti-racism', *Borderlands* 5, no. 3: www. borderlands.net.au/vol5no3_2006/ahmed_nonperform.htm.
Ahmed, S. (2012) *On Being Included: Racism and Diversity in Institutional Life.* Durham and London: Duke University Press.
Amoateng, A. (2015) 'Marriage trends suggest SA is in the racial blender' [Online] Available at: http://city-press.news24.com/Voices/Marriage-trends-suggest-SA-is-in-the-racial-blender-20151002-2 [Accessed 5 July 2017].

Amstutz, M. (2005) 'The promise of reconciliation through truth and some forgiveness in South Africa'. In *The Healing of Nations: The Promise and Limits of Political Forgiveness*, edited by M. Amstutz. Oxford: Rowman and Littlefield Press.

Andruki, M. (2010) 'The visa whiteness machine: transnational motility in post-apartheid South Africa'. *Ethnicities*, 10(3): 358–370.

Applebaum, B. (2008) 'White Privilege/White Complicity: Connecting "Benefiting From" to "Contributing To"'. *Philosophy of Education*, 292–300.

Benson, M. (2011) *The British in Rural France: Lifestyle Migration and the Ongoing Quest for a Better Way of Life.* Manchester: Manchester University Press.

Benson, M. and O'Reilly, K. (2009) 'Migration and the search for a better way of life: a critical exploration of lifestyle migration'. *The Sociological Review*, 608–625.

Caballero, C., Edwards, R. and Smith, D. (2008) 'Cultures of mixing: understanding marriage across ethnicity'. *Twenty-First Century Society*, 3(1): 49–63.

Conway, D. (2009) 'Queering Apartheid: The National Party's 1987 "Gay Rights" Election in Hillbrow'. *Journal of Southern African Studies*, 35(4): 849–863.

Conway, D. (2012) *Masculinities, Militarisation and the End Conscription Campaign: War Resistance in Apartheid South Africa.* Manchester: Manchester University Press.

Conway, D. (2016) 'Shades of white complicity: the end conscription campaign and the politics of white liberal ignorance in South Africa'. In *Exploring Complicity: Concept, Cases and Critique*, edited by A. Afxentiou, R. Dunford and M. New, 119–142. London: Rowan and Littlefield.

Conway, D. (2018) 'Banning taste: boycotts, identity and resistance'. In *The Empire Remains Shop: Cooking Sections* edited by D. Fernandez Pascual and A. Schwabe. New York: Colombia University Press.

Conway, D. and Leonard, P. (2014) *Migration, Space and Transnational Identities: The British in South Africa.* London: Palgrave.

Dubow, S. (2009) 'How British was the British World? The case of South Africa', *Journal of Imperial and Commonwealth History*, 37(1): 1–27.

Du Toit, M. and Quayle, M. (2011) 'Multiracial families and contact theory in South Africa: does direct and extended contact facilitated by multiracial families predict reduced prejudice?'. *South African Journal of Psychology*, 41(4): 540–551.

Edwards, R. (2017) 'Partnered fathers bringing up their mixed/multi-race children: an exploratory comparison of racial projects in Britain and New Zealand'. *Identities: Global Studies in Culture and Power*, 24(2): 177–197.

Fechter, A.-M. (2016) 'Aid work as moral labour'. *Critique of Anthropology*, 36(3): 228–243.

Gilroy, R. (1993) *The Black Atlantic.* London: Verso.

JacobsonC., Amoateng, A. and Heaton, T. (2004) 'Inter-racial marriages in South Africa'. *Journal of Comparative Family Studies*, 35(3): 443–459.

Jaynes, C. (2007) *Interracial Intimate Relationships in Post-Apartheid South Africa* unpublished M.A thesis University of Witwatersrand, Johannesburg.

Leonard, P. (2010) *Expatriate Identities in Postcolonial Organizations: Working Whiteness.* Farnham: Ashgate.

Leonard, P. (2013) 'Making whiteness work in South Africa: a translabour approach'. *Women's Studies International Forum*, 36(Jan–Feb): 75–83.

Leonard, P. Forthcoming. 'Where racism can't be denied: reimagining whiteness and nationhood in contemporary South Africa'. *Identities: Global Studies in Culture and Power.*

Mann, M. (1988) 'The Giant Stirs: South African business in the age of reform'. In *States, Resistance and Cheap in South Africa*, edited by O. Frankel, N. Pines and M. Swilling. Johannesburg: Southern Book Publishers.

McClintock, A. (1995) *Imperial Leather: Race, Gender and Sexuality in the Colonial Context*. New York: Routledge.

McEwan, H. (2013) 'Deserting transformation: heritage, tourism and hegemonic spatiality'. *Diversities*, 15(2): 23–36.

McEwen, H. and Steyn, M. (2013) 'Hegemonic epistemologies in the context of transformation: race, space, and power in one post-apartheid South African town'. *Critical Race and Whiteness Studies*, 9(1): 1–18.

Mills, C.W. (1997) *The Racial Contract*. Ithaca, NY: Cornell University Press.

Omi, M. and Winant, H. (1994) *Racial Formation in the United States: From the 1960s to the 1990s*, 2nd ed. London: Routledge.

O'Reilly, K. (2000) *The British on the Costa del Sol*. London: Routledge.

Schrire, R. (1991) *Adapt or Die: The End of White Politics in South Africa*. New York: Ford Foundation and the Foreign Policy Association.

Song, M. (2016) 'Multi-racial people and their partners in Britain: extending the link between intermarriage and integration?'. *Ethnicities*, 16(4): 631–648.

Steyn, M. (2001) *Whiteness Just Isn't What It Used To Be: White Identity in a Changing South Africa*. New York: SUNY Press.

Steyn, M. (2012) 'The ignorance contract: recollections of apartheid childhoods and the construction of epistemologies of ignorance'. *Identities: Global Studies in Culture and Power*, 19(1): 8–25.

Steyn, M. and Foster, D. (2008) 'Repertoires for talking white: resistant whiteness in post-apartheid'. *South Africa Ethnic and Racial Studies*, 31(1): 25–51.

Steyn, M. and Ballard, R. (2013) 'Diversity and small town spaces in post-apartheid South Africa: an introduction'. *Diversities*, 15(2): 1–5.

Van-Helten, J. and Williams, K. (1983) '"The crying need of South Africa": the emigration of single British women to the Transvaal 1901–1910'. *Journal of Southern African Studies*, 10(1): 17–38.

Walsh, K. (2007) '"It got very debauched, very Dubai!" Heterosexual intimacy amongst single British expatriates', *Social and Cultural Geography*, 8(4): 507–533.

7 British-born Indian second-generation 'return' to India

Priya Khambhaita and Rosalind Willis

Introduction

Half of the British South Asian population were born in the United Kingdom (Peach 2006). Within the South Asian group, just over a half are made up of the Indian population. The Indian migrant population in Britain has aged-in-place, such that second and third generation British Indians now make up a large proportion of this group. We were interested in the relationship between these British-born Indians and their ancestral homeland in the context of increasing emigration. In June 2011, long-term emigration from the UK was 342,000 in total and these figures changed a year later to 352,000 in June 2012 (ONS 2013). There is preliminary evidence that British Indians are indeed exploring India for emigration purposes, whether it is a single long duration of stay, a series of long visits or for permanent settlement. This is not only of academic interest: from the perspective of the UK economy, if this group of expats grows over time and is made up of highly skilled and educated migrants that have strong and extensive professional networks, this could negatively impact on skills shortages and contributions to productivity and growth. From an Indian perspective, however, there may be positives to having a wave of skilled migrants such as these.

Research not only reveals that this migration is taking place but, furthermore, that second or third generation Indians are sending remittances back to their family in the UK (BBC 2012). A land of better economic and investment opportunities is how India is being perceived. It is also seen to have a number of lifestyle benefits with better investment opportunities and improved living standards in terms of accommodation and a wider choice of schools for children (Fernandes 2011). One such example is British-born Mr. Chawla who works as a DJ in a Mumbai nightclub and moved to India to search for better opportunities (BBC News 2012).

This chapter considers second and third[1] generation British-born Indians heading to India. It is a result of a pilot research project.[2] The key research questions were: a) What approaches are return migrants, particularly those in the 'sandwich' generation that are expected to provide care for their children as well as parents and grandparents, adopting to balance their responsibilities? b) In the context of family members living in India and the UK, how do they envisage caring responsibilities to change over coming years? Furthermore, although exploring identity in itself was not a primary research question in this project, the findings of this project are useful for c) exploring motivations to return to India and what can be learnt about the participants' senses of their national identities. A brief discussion of national identities in relation to caring is included here. This chapter is a presentation of our qualitative research findings and extends our knowledge of the prospective migration of British Indian retired elders from the perspective of their adult children and, in the process, what can be learnt about Britishness in this context.

Second-generation return

In recent years the Indian government has reconnected with the diaspora and aimed to strengthen existing bonds to facilitate the diaspora's contribution to the development of modern India and encourage investment in the home country. Their efforts are focused through events set up to celebrate the achievements and contributions of the overseas Indian community, improved ability to travel in and out of India, purchase and invest in property, and have Indian bank accounts (Ministry of External Affairs 2017, Agarwala 2015). Indians across the generations are able to 'return' through the implementation of a political package aiming at strengthening ties with the Indian diaspora. Up to fourth-generation descendants are eligible for the Overseas Citizen of India (OCI) card. This card made it easier to come back, secure a job and stay in India with little administrative hassle. It is a life-long, multiple-entry visa, designed for recently naturalised former Indian citizens and their offspring (Varrel 2010). While facilitating return migration was not a primary goal of the new diaspora policy, migrants have been a side effect.

A sub-group of those returning are not first-generation but second-generation 'returnees'. As these are British-born migrants, they are not returning to their countries of birth as they were not born there and have not lived there. Rather, these are individuals returning to their countries of cultural and religious origin. India is one of the fastest growing economies in the world. Challenging job positions, high demand for experienced workers and economic growth are cited as key factors for creating such reverse talent flows (Jain 2013). In terms of literature on second-generation returnees, preliminary research has been conducted on the second generation in the United States (US) over the past decade, and some of this research interest is now being replicated in Europe (King and Christou 2010). There has been some academic discussion about British second-generation returnees in relation to the Greek, Polish and Caribbean communities (see King and Christou 2010; Reynolds 2008; Górny and

Osipovič 2006). However, our research is unique in that it focuses on second- and third- generation return migration to India.

To our knowledge therefore, this is the first piece of academic research specifically on British-born Indian return migrants. It is important to know about this group of migrants as their flows to and from India may have an impact on social, cultural and economic development in both India and the UK, and by extension will have important implications for British Indian identity. As noted above, the research that does exist has been conducted with US-born Indians and has explored motivations for return and ethnic identities in the workplace (Jain 2013). We are the first to present data on British-born returnees and their caring responsibilities with regards to parents that are resident or 'left-behind' in the UK and what we can in turn learn about constructions of Britishness for this group.

Parents 'left behind' and potential retirement migration

Those 'left behind' can experience various emotions, ranging from emotional ambivalence to anger and distress. Emigration is mostly experienced as a vast loss that brings around significant changes in social networks and relationships (Marchetti-Mercer 2012). The parents of migrants may follow their children and become retirement migrants. They may make 'assistance moves' – where individuals move towards family or other support systems (Conway and Rork 2011). Retirement migration is one of the three distinct types of migration at older ages with the second being health-related migration undertaken to move closer to kin and the third being moves into institutions towards the end of life (Litwak and Longino 1987). We already know that many British elders retire abroad to places such as Spain and South Africa (O'Reilly 2000; Hall 2008; Conway and Leonard 2014). Most research on British migration is on those who identify as White British. This leads to particular perspectives on cultures of retirement and ageing, family, and care. Exploring the experiences of British-Indian families adds to our understanding of different perspectives of these domains. This is in the same way that Seo and Mazumdar's (2011) study highlighted the fragility and fluidity of retirement in the context of migration. In their study of Korean-American seniors, there were elderly parents that moved to the US to be with their adult children, and over time some of them adopted American notions about successful ageing. These include being independent and living separately from family. Similarly, it will be of interest to observe whether, over time, India becomes a retirement destination for British Asian elders retiring to be closer to their migrant children and how this is constructed by retirees.

Methods and sample

This was a small, pilot study and as such involved interviews with four returnees. The first two participants were found via an opportunist approach where they were put in contact with us by friends and family that were aware of the project. Snowballing was then used to recruit the second two participants. All participants were offered an incentive of a $10 Skype airtime voucher. One-off face-to-face semi-structured interviews were conducted in India. Two of these were conducted individually. The remaining two participants expressed a wish to be interviewed together. Interviews ranged from between 45 minutes and an hour and a half in duration. All interviewees gave written informed consent and gave permission for the interviews to be audio recorded.

Details of the sample are given in Table 7.1. Pseudonyms have been used and identifying information has not been included. It is valuable to study a small sample like this for a couple of reasons. First of all, many analytic, inductive, preliminary studies are carried out with small sample sizes. Second, as British Indian returnees to India are yet to be studied in any detail, this small amount of data is important for thinking about where the results might lead in terms of future research priorities in relation to this particular cultural context (Crouch and McKenzie 2006).

The participants were of Punjabi and Gujarati origin respectively. Interviews were carried out in English, however participants commonly included Punjabi and Gujarati words or phrases in their interviews, which Priya understood. Data were analysed using thematic analysis. Key themes include challenges of being present in the lives of transnational families, belonging and extended family living, and cultural sensibilities and adjustment. These themes are discussed in turn in the following sections of the chapter.

Table 7.1 **Participant characteristics**

Participant characteristics	Sukjeet	Hardeep	Shivani	Esha
Sex	Male	Male	Female	Female
Age	43	31	26	24
Education Background	Bachelor's degree	Bachelor's degree	Higher National Certificate	Student – Bachelor's degree
Place of Residence in India	New Delhi	New Delhi	Vadodra, Gujarat	Vadodra, Gujarat
Relationship Status	Married	Engaged	Married	Single
Migration status	Full-time Indian resident	Full-time Indian resident	Full-time Indian resident	Work experience
Parental migration background	East African Indians	Former Indian Nationals	East African Indians	East African Indians
Indian State of Origin	Punjab	Punjab	Gujarat	Gujarat

Findings

Closer proximity and extended family living

Both Shivani and Sukhjeet had grown up in the UK living in an extended family. Sukhjeet's own parents were still living in an extended household in the UK with his younger brother and had grown up in an extended family.

> I actually feel this is more like home than I ever have growing up in the UK where I was surrounded with a lot of love, there was family there, I grew up in a big extended family, I had like my Dad, my three uncles, my three aunties, my grandparents – there was about 26 people up until the age of ten it was a, it was mayhem … .
>
> (Sukhjeet)

Sukhjeet lives in an extended household in India with his wife, daughter and ageing in-laws. Part of the reason he moved to India was that his father-in-law had been diagnosed with cancer, but also in addition to this he had previously worked in India and so was happy to not only be there for his father-in-law but also about the potential to work in India again.

> My father-in-law was diagnosed with cancer, he was 80, yeah he was 80 then, and it wasn't like it was stage four or five, it was stage 2, early stage 2 … You find out, yeah, he's not gonna die tomorrow but we've gotta, it's more psychological for his moral support, ensuring that he follows his diet, takes his medication, do some radiotherapy, diet and exercise. And plus my wife was just worried, she, my mother-in-law is in her late 70s, well early 80s now, so my wife was, it's just, my wife has an elder sister and a brother and then there's her, there's three of them … .

However, he envisages that his own parents may well come to India to live and settle in the future to be closer to him and he has some experience of private health and social care in India due to his father-in-law's condition.

> Thankfully my parents are ok. But if a situation arises with my parents, I'd bring my parents here. There's a better chance for them to get good care, here, living with me. Uhhh, or I'd get a flat nearby for them … it's just you know, with the, it's cheaper, uh you keep an eye on someone here, umm, I don't know, maybe I'm, I'm not really well versed to comment on it 'cause I've never seen the flip side in the UK, thankfully, but if you can find good domestic help, uhh you know, not servants, domestic help, people who work for you.

Leonard's (2008) work highlights how middle-class White British families enjoy a higher standard of living in Hong Kong than at home, including luxury accommodation and domestic help. Not all of the participants here would identify themselves as middle-class; indeed, Sukhjeet self-identifies as being of a working-class background. However, he refers to the same new privileges of domestic help and how these are potential facilitators for providing quality care for elders and making their lives more comfortable.

Care costs and the financial challenges of transnational family life

Sukhjeet's understanding is that availability of quality care is more affordable in India and that this is accentuated by favourable exchange rates and lower living costs. However, he does admit he hasn't had a great deal of experience with services in the UK to make an in-depth comparison. Sukhjeet is one of the three participants whose partners are Indian-born. Therefore, their in-laws live in India and future caring responsibilities are being considered in the context of two sets of parents in two different countries.

In the following quote Hardeep alludes to the financial constraints around being present in the lives and meeting the social and emotional needs of both his parents and in-laws – this would mean a lot of flights between the UK and India supplemented with video-calling and messaging. One concern of his is the potential cost of facilitating this, which is also prompting him to consider changing employers whereby he can earn more money to pay for this care:

> Touch wood they're fairly healthy umm I see caring for parents not necessarily as a cultural thing ... not necessarily completely ... I see it as a human thing right ... To have a lifestyle where I think I can live a life to look after both sets of parents, which means having to be here or there, we have to be able to afford it. Right so this comes down to me maybe leaving my current employer which is not-for-profit. I've been working for them where money was never really like at the front of my mind, it's not necessarily at the front of my mind now but I'm a lot more practical in what I need to get done.

Fulfilling parents' expectations

Esha's views resonate with Sukhjeet's on his parents joining him in India. Esha is an only child. She is a student and she had come to India for one month for work experience as she was training to be an allied health practitioner. Whilst in India she was staying with her relative Shivani who had already made the move from the UK to India. Esha chose India for her work experience placement because she would like to live there one day and thought that the work experience might be a good opportunity to experience day-to-day life in India before she made the final decision to move. Here she discusses her motivations for a potential longer-term migration to India, which are primarily based on concerns about her parents as they approach middle age:

Like my mum's, my mum's job's really, really stressful like she comes home, sometimes she'll come home like 7.30. So it's stressful and they hate it and they're getting like, they're both going to be turning 50 so they're not getting any younger. Like my dad loves India, absolutely loves India and so does my mum, she has family here as well. They're waiting on me to make a move basically 'cause I'm an only daughter too. I've got three years of studying left still. Once I'm qualified and then I can take that step, 'cause I wanna work in Mumbai for at least a year at a hospital. I'm the only child. So my mum's always said like, well in fact I've always said to my mum wherever I go you're gonna come because if I leave London and they stay there then they've got nothing to really stay there for and I don't want them to work in that life up to 70 – up to 60. If I come here they will definitely come over.

There are similarities here with evidence on Turkish German second-generation return migration to Turkey. Kilinc (2014) found that the choice to return was not always an independent decision but was via an initiative of parents. In this scenario, the second generation is fulfilling the expectations of parents who are not yet able to return but have a wish for their children to build their lives in the country of origin.

Furthermore, Shivani's quote below resonates with what Esha and Sukhjeet allude to. She envisages that if she does make a permanent move then her parents will move with her and that the slower pace of life in India would suit her parents as they retire and get older:

Because we've had these conversations and my dad, my dad's very similar to me, and my mum, because we're very simple people, we don't really like living that ... that type of life and coming to India, you can live with very little money but you can have an amazing life, whereas that's just not so possible to do in England because of the expense. My mum's always said that after she retires she wants a very simple life, and she can only get that here. That's definitely something she wants to do, it just depends on her circumstances as to what's going on at that moment.

Constructions of Britishness and Indian identity

In India, multi-generational households are less common than they once were, yet the family still functions as a dominant influence in the life of its individual members (Chadda and Sinha Deb 2013). It may be the case that there is a perception that it is more socially acceptable to do this in India and this, combined with more affordable, culturally relevant quality private health and social care, makes returning to India an attractive option. Even though a multi-generational household is not the exact future envisaged by all of the returnees, India as a collectivist society is being conceptualised as somewhere you can more easily facilitate a close family set up. In a collectivist society there is an emphasis on dependency, empathy, and reciprocity (Chadda and Sinha Deb 2013). The participants discuss their fears about a care gap and voice that they envisage their parents joining them in India. Motivations given for this include a perceived better quality of care and a pace of life and lifestyle perceived to be more suited to healthy ageing.

Unlike with the domains of home and extended family life where these participants identify more with India and being Indian than they do with England and being British, in other areas the returnees identify more with being British. This is in terms of interactions with locals including colleagues, and differences in personal conduct and manners. The quotes below illustrate how the returnees had faced a number of hurdles in adjusting to new ways of communicating and different boundaries around appropriate behaviour:

> It doesn't mean you have to be one of them, but there is a way of talking to people, different types of people, in India to get work done because no one will take you seriously otherwise. That's how India still is. It's not like in London where you email somebody and you can use that as a confirmation to say, right, my work will be done. Over here you have to make 10 phone calls, you have to go see them and say 'is this done?'.
>
> (Shivani)

> I can't get too tactile. Ummm, I've got to watch myself here now, any, any, like if you have a laugh and a joke with a girl in the office in the UK, here they're almost like, my father would be here saying no, you insulted my daughter's honour and you have to go to the *mandir* (temple) now and do *puja* (prayers) for the next ten weeks. So this ... I got into all these scrapes trust me.
>
> (Sukhjeet)

> ... if the British, let's say, have a civil kind of sense there's much less of that here. Ok ... and you have to kind of get used to the fact that people may not have the same, you know, kind of manners, the things that we would find kind of appropriate back in England, you know just a basic thing you know it's hardly a big deal but often here for example people often answer calls in the middle of meetings it's just a bit ... you have to readjust your mind set for what's appropriate.
>
> (Hardeep)

The areas in which these participants faced obstacles included acceptable gestures and greetings between men and women, rules of conduct and courtesy in professional encounters and office culture, and productivity and accountability. These hurdles are in parallel with Reynolds' (2008) second-generation return migrants to the Caribbean. They faced unprecedented difficulties in terms of cultural adjustment and differences in cultural sensibilities.

However, in contrast to Reynold's (2008) findings, the participants in this study did not regard return migration as a short-term measure due to these problems that marked them as different and as 'outsiders'. Instead, only Sukhjeet and Esha expressed vague uncertainty around length of stay in India. This was more related to where they understood themselves in terms of individual career goals and developments in their personal lives rather than cultural adjustment issues.

Discussion

Assistance moves

A key theme of parents re-joining their children is evident amongst these second-generation migrants. When individuals move towards family or other support systems, these are usually called assistance moves that reduce the temporal distance between themselves and their children (Conway and Rork 2011). This theme ties in with what the health secretary had previously said in a speech to the National Children's and Adults Services conference. He explained that he was struck by the 'reverence and respect' for older people in Asian cultures, where it was expected that older grandparents will go to live closer to or live with their children and grandchildren rather than enter a care home (Butler 2013). The Indian group is one of the three ethnic groups in Britain which provide the highest levels of care for a family member with the second and third being White British and White Irish (ONS 2004 cited in Willis 2008).

The cultural value at the core of Asian reverence for older people is still held by the participants. However, they do also highlight that the care for their parents would be brokered with other forms of support and do not expect that they will be able to meet all of the physical, social and emotional needs of their parents simply by living closely to them. In the UK this would primarily mean publicly funded social care, but in India this would equate to privately hired 'good domestic help' as Sukhjeet describes it. Of course, meeting the costs of this is a real concern as highlighted by Hardeep, although favourable exchange rates might ease the pressure of this. This confirms research that formal services are important even where there is a perceived stronger sense of filial responsibility. Even though the cultural value of providing hands-on care for elders might be important, it does not always translate into more comprehensive support (Willis 2008).

Social contract between generations

The 'reverence and respect' for older people in Asian cultures can be understood as a strengthening factor of the social contract between generations. The contract becomes stronger because as children see how their own grandparents are looked after, they develop higher expectations of how they too will be treated when they reach old age (Butler 2013). Sukhjeet's experience evidences the strengthening of the social contract as he wanted to pass down the benefits of living with grandparents that he experienced to his daughter and was happy that she too was living in a multi-generational household.

The Health Secretary has said that if we are to tackle the challenge of an ageing society, we should follow the example of Asian cultures by taking in elderly relatives that move in with them once they can no longer live alone (Butler 2013). This statement was made in the context of funds for NHS and social care services, in particular community services such as home care and day care, being reduced. The difference between private care in India and care available in the UK under the NHS may now be more stark than it might have been previously. These participants might be more worried about ageing parents left behind in the UK in the context of reduced health and social care services than they might be under a fully funded NHS. The scale of impact on budget cuts on such a large scale is difficult to measure. However, the costs of care have undoubtedly shifted on to individuals and their families with a greater reliance and pressure on unpaid family carers (Humphries 2016).

The migrants in this sample have been living multicultural lives with Indian values around reverence and respect in the context of a UK upbringing. They are concerned about the financial implications of being present in their families' lives abroad but at the core of their approaches to transnational caring responsibilities is the hope that their parents will join them in India either close by or in their own household. This is despite the fact that some research has shown that it is possible to maintain mutually supportive relationships across time and space whilst retaining emotional connections intact, particularly when parents benefit from their children's migration through e.g. financial benefits and remittances (Marchetti-Mercer 2012).

Family re-unification in retirement

If these British Asian parents do move to be with their migrant children they would add to the number of British citizens that move abroad to retire. Many Britons retire in areas with a more favourable climate and pace of life. For example, around 100,000 British nationals receive state pension in Spain (Hall 2008). However, this set of parents would be a fascinating set of retirement migrants as they would potentially fall into two further categories: lifestyle migrants and first-generation return migrants. Lifestyle migrants are individuals that migrate either to places that signify, for the migrant, a better quality of life (Benson and O'Reilly 2009). Thirdly they could also be understood as first-generation return migrants if indeed they were born in India, although in this particular sample three out of four participants had at least one East African-born parent.

If the parents were to move to India they might face the challenges of migration in later life and these include cultural and legal barriers which can limit access to and use of support services. Similar to their children, they may face communication, cultural sensibilities and adjustment issues. There may also be cultural differences around expectations of healthcare compared with the UK. These may be further exacerbated by limitations of how the mother-tongue is used day-to-day in present-day India.

Conclusion

This research is the first that includes data on British Indian second-generation return migrants to India. Data is presented on how they envisage their caring responsibilities developing in the context of growing up in an extended household, parents 'left-behind' and new familial links and responsibilities in India via marriage. The BBC recently aired *The Real Marigold Hotel* which followed eight celebrities who wanted to explore the experiences that could be on offer for retirees in India (BBC 2016). The next step after this study is to interview British Indian parents themselves to explore how they envisage their retirement. If these British Indian parents were to move or spend extended periods of time abroad, they might potentially face a number of challenges. These include cultural, language and legal barriers. Where elderly dependents cannot/will not move when they need care, the middle generation will be faced with difficult decisions and financial concerns around meeting needs of two sets of parents in two different countries.

The analysis presented in this chapter has also shed new light on planning for transnational care by British Indians. Previous research has shown that although migrants may wish to remain part of their community through continued contact and remittances, they see migration as a socially acceptable strategy of separation that enables them to keep their own daily lives private outside their community (see Kreager and Schröder-Butterfill 2012). The research presented here indicates that this might not be the case with this particular group. These migrants are keen to have their parents migrate to be close to them to contribute to improving their parents' living conditions and improve their quality of life as they age.

The research discussed here is new evidence on constructions of Britishness by British-born Indians. In relation to the domains of day-to-day social conduct, manners and professional conduct, these returnees are more British than Indian. They subsequently faced cultural adjustment issues. However, in the domains of care, home and family life, the evidence suggests that these participants identify more with India and being Indian than they do with England and being British.

Notes

1 From this point forward, second generation is used as short-hand to refer to both the second and third generation.
2 This project was funded through a Strategic Research Development Fund grant by the University of Southampton.

References

Agarwala, R. (2015) 'Tapping the Indian diaspora for Indian development'. In *The State and the Grassroots: Immigrant Transnational Organizations in Four Continents*, edited by A. Portes and Kelly P. Fernandez Kelly, 84–110. New York: Berghahn Press

BBC (2016) *The Real Marigold Hotel*. [Online] Available at: www.bbc.co.uk/programmes/b06z8840.

BBC News (2012) 'Are you pursuing the Indian Dream?'. [Online] Available at: www.bbc.co.uk/news/18987940.

Benson, M. and O'Reilly, K. (2009) 'Migration and the search for a better way of life: a critical exploration of lifestyle migration'. *The Sociological Review*, 57(4): 608–625.

Boston Consulting Group (2010) cited in Jain, S. (2011) 'The Rights of "Return": ethnic identities in the workplace among second-generation Indian-American professionals in the ancestral homeland'. *Journal of Ethnic and Migration Studies*, 37 (9): 1313–1330.

Butler (2013) 'Jeremy Hunt: UK should adopt Asian culture of caring for the elderly'. *The Guardian*. [Online] Available at: www.theguardian.com/politics/2013/oct/18/jeremy-hunt-uk-families-asia-elderly.

Chadda, R.K. and Sinha Deb, K. (2013) 'Indian family systems, collectivistic society and psychotherapy'. *Indian Journal of Psychiatry*, 55(2).

Conway, D. and Leonard, P. (2014) *Migration, Space and Transnational Identities: The British in South Africa*. London: Palgrave.

Conway, K. and Rork, J. (2011) 'The changing roles of disability, veteran, and socio-economic status in elderly interstate migration'. *Research on Aging*, 33: 256–285.

Crouch, M. and McKenzie, H. (2006) 'The logic of small samples in interview-based qualitative research', *Social Science Information*, 45(4): 483–499.

Fernandes, E. (2011) 'Return of the British Rajas: it's 60 years since the end of the Empire, but thousands of British Indians are heading to the land of their ancestors'. *Daily Mail Online*. [Online] Available at: www.dailymail.co.uk/news/article-2046962/Thousands-British-Indians-heading-land-ancestors.html.

Górny, A. and Osipovič, D. (2006) *Return Migration of Second-generation British Poles*, Centre for Migration Research Working Papers No. 6/64: Warsaw. [Online] Available at: www.migracje.uw.edu.pl/download/publikacja/204/.

Hall, K. (2008) 'The challenges of retiring abroad: a case of older British migrants'. *Regions*, 271: Autumn.

Humphries, R. (2016) 'How serious are the pressures in social care?'. [Online] Available at: www.kingsfund.org.uk/projects/verdict/how-serious-are-pressures-social-care.

Jain, S. (2013) 'For love and money: second-generation Indian-Americans "return" to India'. *Ethnic and Racial Studies*, 36(5): 896–914.

Kilinc, N. (2014) *Second-generation Turkish-Germans Return "home": Gendered Narratives of (Re-)negotiated Identities*. Sussex Centre for Migration Research, Working paper No. 78.

King, R. and Christou, A. (2010) 'Cultural geographies of counter-diasporic migration: perspectives from the study of second-generation "returnees" to Greece', *Population, Space and Place*, 16(2): 103–119.

Kreager, P. and Schröder-Butterfill, E. (2012) *Differential Impacts of Migration on the Family Networks of Older People in Indonesia: A Comparative Analysis.* [Online] Available at: http://core.kmi.open.ac.uk/download/pdf/22861.

Leonard, P. (2008) 'Migrating identities: gender, whiteness and Britishness in postcolonial Hong Kong'. *Gender, Place and Culture*, 15(1): 45–60.

Litwak, E. and Longino, C. (1987) 'Migration patterns among the elderly'. *The Gerontologist*, 7(3): 266–272.

Marchetti-Mercer, M.C. (2012) 'Those easily forgotten: the impact of emigration on those left behind'. *Family Process*, 51: 376–390.

Ministry of External Affairs Government of India (2017) 'Pravasi Bharatiya Divas'. [Online] Available at: https://pbdindia.gov.in/about-pbd.

Office for National Statistics (2004) 'Focus on ethnicity and identity'. [Online] Available at: www.statistics.gov.uk/downloads/theme_compendia/foe2004/Ethnicity.pdf in Willis, R. (2008) 'Older people, ethnicity and social support'. *Generations Review*, 18(4).

Office for National Statistics (2013) *Migration Statistics Quarterly Report.* [Online] Available at: www.ons.gov.uk/ons/dcp171778_300382.pdf.

O'Reilly, K. (2000) *The British on the Costa del Sol.* London: Routledge.

Peach, C. (2006) 'South Asian migration and settlement in Great Britain, 1951–2001'. *Contemporary South Asia*, 15(2): 133–146.

Reynolds, T. (2008) *Ties That Bind: Families, Social Capital and Caribbean Second-Generation Return Migration.* Families and Social Capital Research Group Working Paper, No. 23.

Seo, Y.K. and Mazumdar, S. (2011) 'Feeling at home: Korean Americans in senior public housing'. *Journal of Aging Studies*, 25(3): 233–242.

Varrel, A. (2010) 'Return migration in the light of the new Indian diaspora policy: emerging transnationalism'. In *Dynamics of Indian Emigration. Historical and current Perspectives*, edited by I.S. Rajan and M. Percot, 254–269. New Delhi: Routledge.

Willis, R. (2008) 'Advantageous inequality or disadvantageous equality? Ethnicity and family support among older people in Britain'. *Ethnicity and Inequalities in Health and Social Care*, 1(2): 18–23.

8 Britishness abroad

National identities for contemporary British migrants living in Auckland, Aotearoa New Zealand

Katie Higgins

Introduction

In the last few decades, publications which seek to address identities and lives within the British Isles, at the same time as problematising Britain[1] itself, have seen a remarkable rise in popularity (see, for instance: Colley, 1992; Davies, 2008; Kearney, 1991; Marr, 2000; Nairn, 1981, 2001; Samuel and Light, 1998; Weight, 2002). McCrone (2006, 268) has argued that this willingness to question Britain, even, or especially, as it became a focus of increasing fascination, can be partly traced back to the historian Pocock's (1975) 'plea for a new subject' which addressed each of the four nations, and their relationships with one another, as well as calling for a move beyond an insular island story to explore the global dimension of British history.

With regards to Britain and the dynamics of migration, the focus of most research has been on immigration (see, for example: Burrell, 2006; McGhee, 2009; Modood, 2005). However, a vibrant and growing scholarship has also emerged about contemporary British emigration. This research has addressed the classed dynamics of British emigrants' settlement processes (Benson, 2011; Bott, 2004; Oliver, 2007; Oliver and O'Reilly, 2010; O'Reilly, 2000), the significance of racialised privilege and post/colonial dis/continuities (Coles and Walsh, 2010; Conway and Leonard, 2014; Fechter, 2007; Knowles and Harper, 2009; Leonard, 2010; Schech and Haggis, 2004) and their corporate lives and decision-making as part of a transnational capitalist class (Beaverstock, 2011, 2005, 2002; Cranston, 2016; Harvey, 2012, 2011). Moreover, in Australasia, oral historians have addressed the experiences of post-war, assisted migrants from the UK (Hammerton and Thomson, 2005; Hutching, 1999; Wills, 2005), as well more recent arrivals (Hammerton, 2011, 2010, Pearson, 2014, 2012; Wills and Darian-Smith, 2003). British migrants' diversity with regards to class and gender has been widely acknowledged and explored in this literature. However, this chapter argues that, despite widespread acknowledgement of the contested and heterogeneous character of 'Britishness', there has been relatively little attention to the variation between Britain's constituent nations when researching with them as an emigrant group (for an exception, see: Hammerton and Thomson, 2005). Through an examination of the varied experiences and expressions of 'Britishness' among first-generation migrants in the context of Aotearoa New Zealand, this chapter seeks to address this gap.[2] This attention is crucial for future research with emigrants from Britain to avoid, first, essentialising 'the British' as a homogenous national group, and, second, a related, if unintentional, consequence of this approach, an emphasis upon English experiences and identities, as the majority national culture.

What are the dimensions of 'Britishness' for contemporary, first-generation British migrants living in Auckland, Aotearoa New Zealand? The following analysis is arranged around the experiences of English, Welsh and Scottish participants.[3] First, as has been found in other research (Pearson, 2014), I confirm that English participants tended to distance themselves from overt displays of patriotism and their compatriots who were overly focused on recreating Britain socially and culturally. I expand upon this research through an exploration of some of the ambiguities and contextual specificity of this professed avoidance in practice. Second, I explore the experiences of those who came from nations which might draw on a 'Celtic' identification[4] in order to highlight both the similarities, as well as the significant differences, in their experiences. Finally, I consider the sometimes fleeting and place-bound character of national sentiment.

The pursuit of a 'better Britain' in Aotearoa New Zealand

It is estimated that 22.6 million left the British Isles between 1815 and 1914, in a period which saw 'the expansion of Britain and the peopling and building of the trans-oceanic British world' (Bridge and Fedorowich, 2003: 4, 11). Of those who left Britain and Ireland between 1800 to 1950, only a tiny part, some 500,000, journeyed the 12,000 miles to what became Aotearoa New Zealand (Phillips and Hearn, 2008: vi). In the late eighteenth and early nineteenth century, explorers, missionaries and international traders in flax, timber, seals and whales arrived from across the globe, but from 1788 the relative proximity and vital role British Sydney played in trade meant a distinct British cultural presence developed (McClean, 2012: 11). British colonisation of Aotearoa New Zealand was officially initiated with the signing of the Treaty of Waitangi[5] between the representative of the British Crown, Captain William Hobson, and many, but not all, Māori leaders in 1840 (Orange, 1987). The constitutional evolution of Aotearoa New Zealand has been pragmatic and ad hoc (English and Sharples, 2013: 108–138). The country retains a few lingering constitutional ties with Britain, for instance, Queen Elizabeth II is the current Head of State. However, the fundamental story is one of gradual increasing autonomy (English and Sharples, 2013). Belich (2001) has argued that 1973, when Britain joined the European Economic Community, marked a significant symbolic rift between the countries. However, in terms of immigration status, it was not until the 1987 Immigration Act that British immigrants, 'became non New Zealanders in any "real" sense' (Pearson, 2000: 98). However, 'in a British world that has been dislocated, if not completely lost' (Pearson, 2014: 518), the language, education and skills requirements in recent immigration legislation still give British migrants an advantage over many other nationalities (Pearson, 2013: 86).

As an 'ex-British settler societ[y]' (Anderson, 2000: 382), Pearson and Sedgwick (2010: 447) have suggested that '[s]ettler states like New Zealand formed far-flung nodal points of an empire and can still be placed within a recent increasingly debated postcolonial network emerging out of the British World'. The concept of 'the British World' conveys 'the real and imagined common origins, culture and identity' which connected the globally dispersed sites impacted upon by the British Empire (Bridge and Fedorowich, 2003: 10–11). This network extended beyond the political boundaries of formal empire, and, Bridge and Fedorowich (2003) argue, has a lingering presence after its demise. In addition to this ambiguous context, the challenges within the last few decades to majoritarian national narratives, with the decline of the British Empire, increasing ethnic and cultural diversity, radical economic restructuring and decolonial movements (Pearson, 2008: 52), provides a particularly complex and fascinating site for the exploration of dimensions of 'Britishness' among contemporary arrivals from the UK, a concept unpacked further next.

The 'fuzzy frontier' of Britishness and its constituent nations

In an open and inclusive, if somewhat tautological, approach, (Ward, 2004: 3) has defined Britishness as 'what people mean when they identify themselves individually and collectively as "being British"', which usefully allows for the 'inconsistencies, contradictions and flexibility of daily identity formation'. In another productive framing, Langlands (1999: 63) understands Britishness as an 'added value' and a secondary form of national consciousness, varying in felt intensity according to context. While she concedes that British institutions and public life were largely constructed in English terms, Langlands suggests that participation in them by other nationals does not necessarily conflict with their other identifications (p. 63). Moreover, although many English nationals tend to conflate the two, Langlands challenges the idea 'that Britishness is just Englishness writ large'. Instead, she argues, 'a considerable measure of accommodation, cultural fusion and social intermingling' means the English have been 'Britonised' such that the distinction of England from Britain is 'fuzzier' than it might be for Scottish and Welsh nationals, for instance (p. 63). For many Britons, as Colley (1992: 6) put it, '[i]dentities are not like hats. Human beings can put on several at a time'. In light of the blurred boundaries and intermingling identities of Britishness, next the chapter engages with recent research to have addressed the 'fuzzy frontier' between Britain's constituent nations.

As 'four nations and one' (Kearney, 1991: 4) there are heterogeneous meanings attached to nationality within and between Britain's constituent nations. Williams (2005: 14) speaks of an 'affective borderland' for some

Welsh nationals between England, and a persistent, if minority, adoption of a position of victimhood in relation to the English. In a related dynamic, Condor and Abell (2006) suggest that Scottish nationals in their study found it easier to distance a romanticised patriotism from negative associations to do with the British Empire, than the English (see also: Kiely, McCrone and Bechoffer, 2005; Davidson et al., 2018). Once an imperial nationalism with a global civilising mission, post-empire Englishness has been cast as a national identity that has lost its way (Kumar, 2000: 577). In post-devolution Britain, celebrations of Welshness, Scottishness and Irishness were viewed enviously by the English participants in Clarke, Garner and Gilmour's (2009: 149) study. Many felt that the St. George Cross and Union Jack have become symbols linked to the political right, and were concerned that celebrating Britain and England's past, with its imperial associations, may cause offence. Condor (2000) also found a significant group of English participants were uneasy and embarrassed about expressing what was perceived to be a potentially prejudiced interest in their nationality. In fact, for them, patriotism was equated as tantamount to racism. Finally, Fenton (2007), in his project on young English adults' national sentiment, highlights the importance of including casual indifference, anti-nationalist disregard and a cosmopolitan 'citizen of the world' approach when researching national identities, and more recently has highlighted common positions of 'cosmopolitan-multi-culturalism' or 'resentful nationalism' among the English, themes which are returned to later (Fenton, 2008).

As already seen, there is a tendency for English nationals to have more difficulty separating national pride from British imperial activity in comparison with Scottish or Welsh nationals. The ability to claim an identity of victimhood can offer a valuable, if potentially problematic, position from which to speak. For instance, in a commentary on 'race' and racism in Wales, Williams (1995) has criticised what she calls 'the tolerance thesis'. She argues that 'Welsh people's claim to an understanding and empathy with oppression' (p. 119) can lead to an inability to reflect on racism in Wales, which is viewed as an English problem, following the logic that 'oppressed peoples cannot be oppressors' (p. 120). It appears that a difference in the meaning of patriotism between Britain's constituent nations can be found in national 'descendants' beyond Britain's borders too. In an exploration of ancestral heritage-tourism in Scotland, Basu (2005: 147) has explored the attraction for the 'morally dispossessed' descendants of settlers in ex-British settler societies to participate in a collective 'Celtic dreaming', which casts them as victims rather than perpetrators of displacement (see also: Curthoys, 1999). In Aotearoa New Zealand, Paterson (2012) has suggested that English ancestry is felt to be awkward for Pākehā New Zealanders[6] reclaiming their migrant origins due to a stronger association of England with the British Empire. Nationals from all over Britain and Ireland were involved in the thousands of individual actions and intentions that made up imperialism, and effective power quickly devolved to settler elites in Aotearoa New Zealand. Yet, Paterson (2012) suggests, however naively, that those of Catholic Irish background, to a limited extent the Scottish, and I would add the Welsh, may 'reclaim an association with their ancestral homelands with slightly less postmodern angst than those of purely English origins' (p. 125). The lack of ethnic or regional societies for the English in Aotearoa New Zealand historically and today, in distinction to the continuing success of Scottish Associations (Harper, 2011: 234), for instance, may also be partly attributed to their forming a central aspect of the majority Pākehā culture (Paterson, 2012: 124). In this chapter I examine the differences and similarities in the meaning of national identities and expressions of patriotism for contemporary English, Welsh and Scottish migrants to Aotearoa New Zealand.

Research methods

This chapter draws on a 12-month qualitative study in Auckland with first-generation British migrants from May 2013 to April 2014. The broader project explored British nationals' migration stories, national belonging, personal geographies and reflections on Auckland's ethnic and cultural landscapes. I conducted in-depth interviews with 46 British migrants, as well as speaking to an eclectic group of local 'key actors'. A smaller group of participants, chosen to best reflect a heterogeneity of experiences, participated in an ongoing series of research encounters, where I would join them walking and driving around places which were significant to them, as well as requesting that they photograph a week in their everyday lives and places of Auckland. Finally, I spent time with participants informally in their everyday lives, visiting their homes, attending events and places in their neighbourhoods, as well as making repeat observations at British-themed commercial establishments, events and societies.

Aotearoa New Zealand was the seventh most popular emigration destination for Britons in 2008 with 248,000 nationals living there long-term (Finch et al., 2010: 29). In contrast with other migrant groups, most British migrants live outside of the Auckland region. However, Auckland has the largest concentration of British migrants, with over 90,000 people identifying the UK as their birthplace in 2013 (Statistics New Zealand, 2013a). British migrants are scattered throughout the city, but they tend to be concentrated in affluent coastal suburbs (Gilbertson and Meares, 2013). In order to engage with this large and relatively dispersed group, I recruited participants through placing calls in suburbs with a high concentration of British nationals and subsequent 'snowballing'. This stage was supplemented by recruiting people I met in my everyday life and placing advertisements on 'British-expat' online forums and posters in British-themed shops and pubs. Although the latter approach may seemingly encourage a bias to nationally oriented respondents, this recruitment strategy only recruited a small minority of participants. I actively sought to include participants who felt ambivalent about being 'British' through an eclectic approach to recruitment at different events, places and through various social networks. However, when I tried to recruit members of the Auckland Irish Society, the inclusion of 'the British' in my recruitment material was off-putting for the Northern Irish people I spoke with, who, understandably, identified as Irish instead. Aside from a flaw in my research design, this highlighted some of the difficulties of working with such a contested national category.

In terms of participants' characteristics, all four of Britain's constituent nations participated. However, the predominant nationality within 'the British' in Aotearoa New Zealand is English (Statistics New Zealand, 2013b) and that was also the predominant nationality among those I researched with. Three participants were people of colour, but the majority were racialised as white. Rather than only include those who had been in residence for a particular length of time, I asked that all participants have a New Zealand permanent residency visa or citizenship, the difference between which is minimal in terms of rights (Spoonley and Bedford, 2012), as opposed to a working holiday or tourist visa, for instance. The length of residence among participants ranged from 6 months to 56 years. However, the majority were long-term residents with an average length of residency of 16 years. Those who travelled prior to the 1980s reforms were more likely to identify as working class, although many had experienced social mobility since then (Pearson, 2013). However, most had arrived after a series of immigration reforms in the 1980s which necessitated migrants have specific occupations or skills. While class is dynamic, relational and contingently experienced, the largely professional occupations and apparent affluence of the majority of participants meant that they could be described as middle class. Finally, the participants in the study ranged in age from those in their early twenties to their late eighties with a median age of 51.

My embodiment shaped the research process. I am English, British, white and middle class which ostensibly positioned me as an 'insider' with many participants in this study. However, insider-ness and outsider-ness is better thought of as an ongoing, unfolding of proximity and distance composed of 'momentary spaces' (Mullings, 1999: 340). Nowicka and Cieslik (2014) challenge the latent understanding that common origin produces 'common individuals'. In fact, social proximity, paradoxically, can increase awareness of social divisions, such as class or generation (Ganga and Scott, 2008). The ways in which participants related to me as an insider/outsider – for instance, whether assuming complicity between 'us' about 'them' or alternately distancing themselves from my Englishness and middle-class status – provided insights about how they made sense of who they were and how they represented themselves as part of a community (Young, 2004: 200).

I adopted an iterative-inductive approach to analysis (O'Reilly, 2005), returning to the material gathered again and again to draw out recurrent patterns and more idiosyncratic events in ongoing conversation with previous empirical research and concepts drawn from the wider literature. The following analysis does not claim to be representative, instead aiming to convey some of the complexities and contestations of 'British' national identities among participants in this study. While informed by the experiences of the larger group of research participants, this chapter highlights five participants' experiences who illustrate key themes with regards to the dimensions of Britishness observed among this group. They articulate the ambiguities of a general professed avoidance of overt displays of patriotism among English/British participants, the cultural capital attached to 'Celtic' identities and the sometimes fleeting or place-bound emergence of patriotic sentiment.

Whingeing Poms: Englishness and the avoidance of displays of patriotism

How can you tell when a plane full of Poms has just landed?
The whining carries on after the engines have been switched off.

The characteristics of 'the whingeing Pom', a figure evoked in the locally well-worn opening joke above, is associated with arrivals from the UK with a superior attitude and a propensity to complain.[7] This figure, and their associated behaviour, was negatively perceived by most participants in this study (see, also: Pearson, 2014: 514). Moreover, several Welsh and Scottish participants distanced themselves from this label altogether by claiming it was associated with the English. In addition to the derided figure of 'the whingeing Pom', in Pākehā society the notion of 'cultural cringe' provides important context. A legacy of Aotearoa New Zealand's history as a British colony, this term describes 'an insecure attitude to local culture' in relation to other cultures and, especially, historically in relation to Britain (Horrocks, 2004: 280 cited by Barnes, 2012: 270). This phenomenon can be connected with a particular local sensitivity to overbearing or superior performances of Britishness, and, more specifically, Englishness. For contemporary English migrants in Aotearoa New Zealand, my research confirmed that overt signs of nationness are rare, and visible signs of 'waving the flag' and 'vulgar displays abroad' are condemned by men and women of varied ages, length of residence and occupations (Pearson, 2014: 513). The 'bad migrant' who failed to integrate was often a classed figure. However, attention to some of the ambiguities of English migrants' professed avoidance of patriotism points to a more nuanced picture.

In his early fifties, Martin[8] first travelled to Aotearoa New Zealand from rural Somerset in England, and the house he had been born in, with his wife, who was Scottish. A few weeks into their visit they decided to migrate and he had been living in Auckland for three years when I met him. Martin was dismissive of what he considered typical 'Brits abroad' who are 'mouthy and loud'. Instead, he wanted to 'blend in' and be an 'honorary Kiwi'. When I asked him to expand on what 'blending in' might involve, he answered,

I don't go around with England t-shirts and England hats on and Union Jacks plastered everywhere ... I don't go out of my way to be English.

However, Martin illustrated some of the ambiguity of integration among participants as he lived in a neighbourhood popular with British migrants and was a regular at an English ale house which he described as 'my local at the moment'. He would often drink there at the weekend, watch the sun set and call his family and friends in the UK. Later in our conversation, Martin reflected further upon his attachment to his nationality.

> I, kind of, get the odd times I've heard the national anthem since I've been in New Zealand I, kind of, get a bit patriotic, you know, not *emotional*, but I suppose a bit patriotic, you know. It doesn't matter how hard I try to be a Kiwi, I'm still gonna be an English man, so– and I'm not ashamed of that, um, I suppose the thing I am ashamed of is the way England has gone down the tubes. From the Great– I don't– England was the greatest nation on Earth at one time … Am I patriotic? Yeah, I am, yeah, I am, but I still wouldn't be– if they had a, uhh, if they had a– what's gonna happen now, before long they're gonna have a christening for the Royal sprog [baby], aren't they? I mean there'll probably be a do down here somewhere, but I won't show up for it.

Martin's opinion fits into a broader 'narrative of decline' tracked among the English in the UK (Fenton, 2008; Clarke et al., 2009). His ongoing national attachment can be framed as what Fenton (2008) has described as resentful nationalism, in which exclusion from a multicultural national discourse, nostalgia for a more homogenous 'golden age' and a declining state which fails to uphold certain aspects of civic life, undermines the ability to identify positively with the nation. Although Martin distanced himself from 'overly' patriotic or nationally oriented compatriots, in a pattern which could be tracked across the majority of participants in this study, he also points to more of an ongoing attachment to his nationality than his initial claim to distance might suggest.

Moreover, through claiming a playful, ironic position, a minority of English participants did engage in 'flag-waving'. In Henry's case, literally. He was in his late forties and came to Auckland with his Pākehā wife and their three children seventeen 17 years previously, leaving behind a one-bedroom flat in South London, England. He had decorated a space at his work, where he and some other English migrants had been placed together, with Union Jack flags, which he explained to me, repeatedly, were 'a little ironic'. Henry then went on to add, 'there's something about being English in New Zealand that feels funny'. Henry expressed frustration that the English were singled out by some Pākehā New Zealanders as colonisers in a way which erased their own complicity. He explained what his national identity meant to him,

> as an English person, regardless of whether I– I'm not some Colonel Blimp or anything, but until you come away you don't realise that, you know, I *am* English, and I grew up in England and learned all the things

about being English. And when you go away from there you realise, yeah, they're actually part of you and if someone doesn't respect those they're not respecting you. Much like anything that really is part of your character, I suppose ... I'm very much, kind of, into neuroscience and that, sort of, you know, that's what you develop. That's how your brain develops when you grow up. It's not entirely hard wired but it's pretty much– by the time you're an adult therefore that's what– that's what you are.

Henry stressed that the flags at work were ironic and distanced himself from the jingoistic figure of 'Colonel Blimp', a pompous, stereotypical British cartoon character first created, by a New Zealander, for the UK newspaper the London Evening Standard. However, in response to a perceived sense of animosity from Pākehā towards the English, his national attachment was conveyed as almost irresistible, in a biological explanation of a national habitus (Edensor, 2002: 93–95). Bourdieu's (1987) reflection on the phenomenon of 'slumming it' offers an insight into his ironic patriotic behaviour. Bourdieu suggested that when intellectuals or artists read popular novels or watch Westerns, they could transform such works into props of distinction through distancing or ironic readings which are thus still governed by the organising principles of the bourgeois aesthetic habitus (referenced by Bennett et al., 2009: 26). Even as Henry's ostensibly overt displays of patriotism illustrated an exception to its general avoidance among British and, especially, English participants, through his self-consciously ironic performance, he can be argued to reaffirm that principle.

The migrant who was open to new horizons was a highly valued subject position among many participants in this study. In contrast, compatriots who were overly focused on recreating Britain socially and culturally, and had thus failed to integrate, were generally viewed as 'bad migrants' among participants from across Britain (see, also: Benson, 2011; O'Reilly, 2000). However, the examples outlined above highlight a tendency for a stronger ongoing relationship with their Englishness/Britishness among participants than their initial rejection suggests. Participants were able to engage in British cultural and social activities, at the same time as distancing themselves from them. Although it can be overstated, this pattern reflects the privileges of ongoing cultural similarity with the majority Pākehā culture. Pubs and neighbourhoods popular with the British, for instance, were also generally popular with Pākehā New Zealanders. Their cultural and social activities tended to be viewed as 'ordinary' rather than 'ethnic' by the majority culture. However, 'the British' have varying relationships to their nationality, as will be revealed next through an exploration of participants' experiences from Wales and Scotland.

Celtic capital: the differing experiences of Welsh and Scottish migrants

In his early seventies, David had grown up in a small town in the Swansea Valley in Wales. He met his partner, who was Pākehā, while they were both working in London. In the 1980s he was made redundant and they decided to trial living in Auckland. David described himself as 'a Welsh man full stop and a Brit if you like as well. I can certainly relate to being a Brit and a Welsh man'. Aside from his accent, he displayed his nationality visibly. David had the Welsh flag on a bumper sticker on his car, wore a hat with the red dragon when we went to an amateur British tournament rugby match together and was an active member of the Welsh Society in Auckland, which was how I met him. David experienced some of the same discomfort with historical colonial relations as his British compatriots. For instance, he spoke of 'keeping quiet' during Treaty training sessions at work because of Welsh missionary involvement in the colonial era.[9] However, when he was called a Pākehā, a potentially politicised identity (Spoonley, 1991), he told me, 'I say, "no I'm a Taff!"' repeating the rhyme, 'Taffy was a Welshman, Taffy was a thief, Taffy went to the butcher's shop and stole a leg of beef!' In this way, David claimed a more ambiguous identity than that of a settler in Aotearoa New Zealand through stressing his Welsh heritage. Although perhaps fondly appropriated, rather than a position of privilege the rhyme points to a history of cultural marginalisation for Welsh people in the UK. The most overt, and least self-conscious, displays of patriotism I saw were from Welsh nationals. This trend may reflect my access to the Welsh Society and the Welsh Club. However, I want to argue that it also indicates different associations with 'Celtic' nationness. Not all of the Welsh and Scottish participants I spoke to were as vocal in their patriotism as David, and I turn next to less nationally oriented examples.

I met Charles when he responded to a poster I had put up in his suburb of Devonport, a coastal neighbourhood popular with British migrants. He was in his late fifties and from Cardiff in Wales. However, he was educated in private boarding schools around the UK, where he had developed a Home Counties English accent. He migrated with his wife, who was English, 34 years previously. He usually identified as British, because of a perception of its greater inclusivity, telling me, 'I think British is a nice expression because it, sort of, it almost captures the essence of multiculturalism'. Charles opposed this identity to his distaste for 'huge nationalism, you know ... in certain times in the UK you've got lots of, sort of, the St George Cross and whatever, wherever you go'. In this way, Charles fits Fenton's (2008) description of the 'cosmopolitan-multiculturalist'. This figure occupies a position of relative detachment from their nationality and is associated with a global elite, intelligentsia and ethnic minorities in the UK. However, he told me he was 'quite proud if anyone calls me Welsh'. In contrast to his distaste for the English flag,

> actually when we go sailing we quite often – we've got a family flotilla – we put a huge Welsh flag up on the mast. An enormous great big Welsh flag.

Although not all participants from Wales engaged in such patriotic displays, the contrast between Charles' perception of the English and Welsh flag usefully highlights the differences in meaning attached to overt displays of patriotism between the nations. These differences in meaning can be connected with distinct, if connected, national histories and the possibility of claiming a more romanticised patriotism.

As mentioned, not all Welsh and Scottish nationals engaged in such overt displays of patriotism. As a teenager growing up in Edinburgh, Scotland, Aileen had developed a passion for languages after a school trip to Paris, France. She later travelled around Europe and South America before she met her partner, who was Pākehā, while living in Australia. They moved to Auckland almost thirty years ago after they had their first child. Now in her late fifties, I met her through another participant who lived in the same suburb of Titirangi. When I asked if she had heard that her neighbourhood was sometimes referred to as 'Britirangi' because of its popularity with British migrants, she replied, 'No, really? Because there's so many British people? Oh, that puts me off actually. I don't gravitate. I tend to go the opposite way'. She continued,

> if I hear a Scot I'm not– I don't gravitate towards them necessarily. But one of the interesting things I think is that, and you probably experienced it as well, there's a kind of unspoken knowledge that you share with somebody from your own heritage, background. Like I work with a Scot, he's from Glasgow, and there's certain things you don't need to explain, uh, with each other 'cos you're both, kind of, from the same background. You instantly develop a kind of unspoken relationship, if you like. That's what I find with the Scots ... Probably more so Scottish 'cos there's that– there's that tension ... between the Scottish and the English, because of the history and the culture and the politics and everything ... I mean it's in the history and it's in the attitude, you know, [with a sneer] the [fuc]kin' English, you know, 'cos that's what the Scots think, because of the Jacobite rebellion and it goes way back to before the seventeenth century it's really deep ... there's a lot of prejudice towards the English.

Aileen articulates an unspoken, tacit understanding between compatriots through her encounters as a migrant: a vague sense of the possibility for shared understandings, cultural references and humour. However, Aileen spells out different layers of identification within 'the British' and her sense of relative distance from the English which she attributes, among other things, to histories of enmity.

While she distanced herself from patriotism and nationally oriented pursuits, in the next extract, Aileen describes a fleeting but vivid nostalgic connection with her Scottishness. She told me about attending a conference in Dunedin, Aotearoa New Zealand where, as part of a group, they drank whisky, visited a Robert Burns statue and were offered a reading of his poems, which she then volunteered to perform.

> ...I just went straight into this Scottish accent and read this poem and I got really, really, kind of, emotionally involved in this poem. And

everybody was like, 'oh my god! What are you doing?' And this woman just went, 'that was amazing!' I was like, 'I know'. I don't know what came over me. I just was, like, right in that poem underneath the Robbie Burns statue. That's the only time really in New Zealand, I think, I've, kind of, wanted to identify and portray myself as 100% Scot ... and so– and that really surprised me, and I got quite emotional about it. I was almost in tears reading this poem.

The ability to express national attachment was differentiated for Britons in Aotearoa New Zealand. Aileen's ability to proudly claim and perform her Scottish heritage is more difficult to imagine for an English migrant. National sentiment is realised through spectacular rituals and displays (Hobsbawm and Ranger, 2012), everyday, banal iterations (Billig, 1995; Edensor, 2002; Fox and Miller-Idriss, 2008) and dynamic, situational configurations of 'affective atmospheres' (Closs Stephens, 2015). Closs Stephens (2015) has explored national feelings and the way they take hold through 'affective atmospheres' in an attempt to circumvent 'the language of identity, essence and belonging' in research on nationality. Instead, she attends to 'the currents and transmissions that pass between bodies and which congeal around particular objects, materials and bodies in specific times and spaces' (p. 12). The heady mix of place – a Scottish settlement, Dunedin is the Gaelic word for Edinburgh and the two cities share street names – whisky, a national icon's statue and poetry, all combined to elicit a brief but strong sense of national attachment for Aileen, despite herself. Her experience is reminiscent of Martin's sudden rush of patriotic feeling when he heard the national anthem. These examples illustrate the way in which national attachment is more contingent and dynamic than reflections upon identities might immediately account for.

Conclusion

This chapter examined dimensions of Britishness among contemporary first-generation migrants in Auckland, Aotearoa New Zealand. It complemented previous research which has highlighted ambivalence about overt performances of Britishness internationally for British emigrants (see, for example: Benson, 2011; Coles and Walsh, 2010; Hammerton, 2011; Oliver and O'Reilly, 2010; Pearson, 2014). However, it sought to expand this research through attention to some of the differences, and commonalities, between the various nationals of Britain's constituent nations. This focus on national identities was intended to draw out in greater detail an important aspect of the heterogeneity of this emigrant group otherwise neglected in this field of research.

British emigrants' experiences take place within, and are influenced by, a broader context of shifting representations of Britishness. However, their nationality is contingently realised in different places according to the specific historical, political, social and cultural context. Overt expressions of patriotism were generally felt to be less acceptable for the English, in comparison to Welsh and Scottish, participants, in this study. English nationals' experiences can be connected with a specific regional reputation for 'whingeing Poms', the significance of the notion of 'the cultural cringe' for Pākehā society, as well as a widely held association between this national identity and the British Empire. Welsh and Scottish nationals' relationships with displays of patriotism evidenced significant differences. Although many shared a common distancing from 'flag-waving' or what was perceived to be a 'narrow-minded' interest in recreating their national lives socially and culturally, I tracked a greater ability to celebrate their nationality without the negative associations outlined for the English. As well as a rise in cultural capital attached to romanticised Celtic identities which has been documented internationally (Harvey et al., 2002: 14), this pattern can be connected with a locally inflected desire for a more 'innocent' ancestral heritage among some Pākehā and different historical associations for the 'imperial' English and 'Celtic' Scottish, Welsh and Irish nations of Britain. Attention to the dimensions of Britishness among emigrants in nations shaped by histories of British colonial expansion and contemporary mobilities offers an interesting avenue for future research.

Notes

1   In order to situate my use of this term and, in particular, its remit, I want to quote Kumar's (2000: 595) clarifying discussion of 'Britain' at length.

> 'Britain' – properly 'Great Britain' – is the political entity formed by the Union of English and Scottish crowns (1603) and parliaments (1707): it followed an earlier annexation of Wales by England (1536). It became the 'United Kingdom of Great Britain and Ireland' by the union with Ireland in 1801; and the 'United Kingdom of Great Britain and Northern Ireland' on the formation of the Irish Free State … in 1921. Strictly speaking … 'Britain' and 'British' exclude the Irish of Ulster (Northern Ireland), through that has not stopped both British and Northern Irish from using the terms as if they included both groups.

2   Regional, classed, aged, ethnicised, racialised and gendered identities, inter alia, intersected with these national variations, but they have not formed the central focus of this chapter due to limitations of space.

3   As I will expand upon in the discussion of research methods, I had trouble accessing participants who identified as Irish. Two Northern Irish nationals took part in this study. However, they identified as British. I have excluded the Northern Irish experience from this chapter because of a lack of perspective in this study rather than a lack of relevance.

4   I use this term bearing in mind McCrone's (2006: 270) criticism that the use of 'Celts' can attribute 'the rise of nationalism in these territories' as 'the outcome of inherent, genetic differences rather than a reaction to politico-cultural processes'.

5   There were several versions of this Treaty, Hobson signed a version in English; whereas most Māori signed the versions in te reo, the Māori language. There were crucial differences in translation between the different versions of the Treaty which are an ongoing source of conflict and debate.

6 Pākehā is used to refer to New Zealanders of European descent. However, it is worth acknowledging that this term is contested, and can obscure differences within this group.
7 Wellings (2011) claims that Pom, which can be affectionate or derogatory, is an abbreviation of 'pomegranate' which plays on the word 'immigrant' and was used to refer to the reddish complexion of new, sunburnt arrivals. The label is also linked with demonised post-war British migrants involved with unionism, labelled the 'British disease', and another popular origin tale suggests that it is an acronym of 'prisoners of her majesty' or 'mother England' in reference to the historic forced transportation of convicts from Britain to Australia.
8 All names are fictitious.
9 As part of national efforts to adopt a bicultural approach to public services which aims to acknowledge Māori, the indigenous people of Aotearoa New Zealand, and Pākehā. Many employers offer training in bicultural awareness and cultural responsiveness, which includes learning about the Treaty of Waitangi.

References

Anderson, K. 2000. 'Thinking "Postnationally": Dialogue across Multicultural, Indigenous, and Settler Spaces'. *Annals of the Association of American Geographers*, 90: 381–391.

Barnes, F. 2012. *New Zealand's London: A Colony and its Metropolis*. Auckland: Auckland University Press.

Basu, P. 2005. 'Macpherson Country: Genealogical Identities, Spatial Histories and the Scottish Diasporic Landscape'. *Cultural Geographies*, 12: 123–150.

Beaverstock, J. 2002. 'Transnational Elites in Global Cities: British Expatriates in Singapore's Financial District'. *Geoforum*, 33: 525–538.

Beaverstock, J. 2005. 'Transnational Elites in the City: British Highly Skilled Inter-Company Transferees in New York City's Financial District'. *Journal of Ethnic and Migration Studies*, 31(2): 245–268.

Belich, J. 2001. *Paradise Reforged: A History of the New Zealanders*. Honolulu: University of Hawai'i Press.

Bennett, T., Savage, M., Warde, A., Gayo-Cal, M. and Wright, D. 2009. *Culture, Class, Distinction*. London: Routledge.

Benson, M. 2011. *The British in Rural France: Lifestyle Migration and the Ongoing Quest for a Better Way of Life*. Manchester: Manchester University Press.

Billig, M. 1995. *Banal Nationalism*. London: Sage.

Bott, E. 2004. 'Working a Working-Class Utopia: Marking Young Britons in Tenerife on the New Map of European Migration'. *Journal of Contemporary European Studies*, 12: 57–70.

Bourdieu, P. 1984. *Distinction*. Abingdon:Routledge.

Bridge, C. and Fedorowich, K. 2003. 'Mapping the British World'. *The Journal of Imperial and Commonwealth History*, 31: 1–15.

Burrell, K. 2006. *Moving Lives: Narratives of Nation and Migration among Europeans in Post-War Britain*. Ashgate: Aldershot.

Clarke, S., Garner, S. and Gilmour, R. 2009. 'Imagining the Other/Figuring Encounter: White English Middle-Class and Working-Class Identifications'. In *Identity in the Twenty-First Century: New Trends in Changing Times* edited by M. Wetherell, 139–156. Basingstoke: Palgrave Macmillan.

Closs Stephens, A. 2015. 'The Affective Atmospheres of Nationalism'. *Cultural Geographies*, 1–18.

Cohen, R. 1995. 'Fuzzy Frontiers of Identity: The British Case'. *Social Identities*, 1: 35–62.

Coles, A. and Walsh, K. 2010. 'From "Trucial State" to "Postcolonial" City? The Imaginative Geographies of British Expatriates in Dubai'. *Journal of Ethnic and Migration Studies*, 36: 1317–1333.

Colley, L. 1992. *Britons: Forging the Nation 1707–1837*. New Haven: Yale University Press.

Condor, S. 2000. 'Pride and Prejudice: Identity Management in English People's Talk About "This Country"'. *Discourse and Society*, 11: 175–205.

Condor, S. and Abell, J. 2006. 'Romantic Scotland, Tragic England, Ambiguous Britain: Constructions of "the Empire" in Post-Devolution National Accounting'. *Nations and Nationalism*, 12: 453–472.

Conway, D. and Leonard, P. 2014. *Migration, Space and Transnational Identities: The British in South Africa*. Basingstoke: Palgrave Macmillan.

Cranston, S. 2016. 'Producing Migrant Encounter: Learning to be a British Expatriate in Singapore through the Global Mobility Industry'. *Environment and Planning D: Society and Space*, 34: 655–671.

Curthoys, A. 1999. 'Expulsion, Exodus and Exile in White Australian Historical Mythology'. *Journal of Australian Studies*, 23: 1–19.

Davidson, N., Liinpää, M., McBride, M. and Virdee, S. 2018. *No Problem Here: Racism in Scotland*. Edinburgh: Luath Press Ltd.

Davies, N. 2008. *The Isles: A History*. London: Pan Macmillan.

Edensor, T. 2002. *National Identity, Popular Culture and Everyday Life*. Oxford: Berg.

English, B. and Sharples, P. 2013. *New Zealand's Constitution: A Report on a Conversation*. Wellington: Ministry of Justice.

Fechter, A.-M. 2007. 'Living in a Bubble: Expatriates' Transnational Spaces'. In *Going First Class? New Approaches to Privileged Travel and Movement* edited by V. Amit, 33–52. New York: Berghahn Books.

Fenton, S. 2007. 'Indifference Towards National Identity: What Young Adults Think About Being English and British'. *Nations and Nationalism*, 13: 321–339.

Fenton, S. 2008. 'The Semi-Detached Nation: Post-Nationalism and Britain'. *Cycnos*, 25: 245–260.

Finch, T., Andrew, H. and Latorre, M. 2010. *Global Brit: Making the Most of the British Diaspora*. London: Institute of Public Policy Research.

Fox, J. and Miller-Idriss, C. 2008. 'Everyday Nationhood'. *Ethnicities*, 8: 536–576.

Ganga, D. and Scott, S. 2008. 'Cultural "insiders" and the issue of positionality in qualitative migration research: moving "across" and moving "along" researcher-participant divides'. *Forum: Qualitative Social Research*, 7(3): 1–12.

Gilbertson, A. and Meares, C. 2013. *Ethnicity and Migration in Auckland*. Auckland: Auckland Council Technical Report.

Hammerton, J. 2011. '"Thatcher's Refugees": Shifting Identities among Late Twentieth-Century British Emigrants'. In *Britishness, Identity and Citizenship: The View from Abroad* edited by C. McGlynn, A. Mycock and A. McAuley, 233–248. Oxford: Peter Lang.

Hammerton, J. 2010. '"Growing Up in White Bread England in the 60s I Might as Well Have Come from Mars": Cosmopolitanism and the City in the Lives of British Migrants in the Late Twentieth Century'. *History Australia*, 7: 32. 1–32. 11.

Hammerton, J. and Thomson, A. 2005. *Ten Pound Poms: Australia's Invisible Migrants*. Manchester: Manchester University Press.

Harper, M. 2011. 'A Century of Scottish Emigration to New Zealand'. *Immigrants and Minorities: Historical Studies in Ethnicity, Migration and Diaspora*, 29: 220–239.

Harvey, D., Jones, R., McInroy, N. and Milligan, C. eds. 2002. *Celtic Geographies: Old Culture, New Times*. London: Routledge.

Harvey, W. 2011. 'British and Indian scientists moving to the US'. *Work and Occupation*, 38: 68–100.

Harvey, W. 2012. 'Labour Market Experiences of Skilled Migrants in Vancouver'. *Employee Relations*, 34: 658–669.

Hobsbawm, E.J. and Ranger, T.O. eds. 2012. *The Invention of Tradition*. Cambridge: Cambridge University Press.

Hutching, P. 1999. *Long Journey for Seven Pence*. Wellington: Victoria University Press.

Kearney, H. 1991. *The British Isles: A History of Four Nations*. Cambridge: Cambridge University Press.

Kiely, R., McCrone, D. and Bechoffer, F. 2005. 'Whither Britishness? English and Scottish People in Scotland'. *Nations and Nationalism*, 11: 65–82.

Knowles, C., and Harper, D. 2009. *Hong Kong: Migrant Lives, Landscapes and Journeys*. Chicago: University of Chicago Press.

Kumar, K. 2000. 'Nation and Empire: English and British National Identity in Comparative Perspective'. *Theory and Society*, 29: 575–608.

Langlands, R. 1999. 'Britishness or Englishness? The Historical Problem of National Identity in Britain'. *Nations and Nationalism*, 5: 53–69.

Leonard, P. 2010. *Expatriate Identities in Postcolonial Organisations: Working Whiteness*. Farnham: Ashgate.

Marr, A. 2000. *The Day Britain Died*. London: Profile Books.

McClean, R. 2012. 'Introduction'. In *Counting Stories, Moving Ethnicities: Studies from Aotearoa New Zealand* edited by R. McClean, B. Patterson and D. Swain, 1–26. Hamilton: University of Waikato.

McCrone, D. 2006. 'A Nation That Dares Not Speak Its Name? The English Question'. *Ethnicities*, 6: 267–278.

McGhee, D. 2009. 'The Paths to Citizenship: A Critical Examination of Immigration Policy in Britain Since 2001'. *Patterns of Prejudice*, 43(1): 41–64.

Modood, T. 2005. *Multicultural Politics: Racism, Ethnicity and Muslims in Britain*. Edinburgh: Edinburgh University Press.

Mullings, B. 1999. 'Insider or Outsider, Both or Neither: Some Dilemmas of Interviewing in a Cross-Cultural Setting'. *Geoforum*, 30: 337–350.

Nairn, T. 1981. *The Break-Up of Britain: Crisis and Neonationalism*. London: Verso.

Nairn, T. 2001. *After Britain: New Labour and the Return of Scotland*. London: Granta Books.

Nowicka, M. and Cieslik, A. 2014. 'Beyond Methodological Nationalism in Insider Research with Migrants'. *Migration Studies*, 2(1): 1–15.

Oliver, C. 2007. *Retirement Migration: Paradoxes of Ageing*. New York: Routledge.

Oliver, C. and O'Reilly, K. 2010. 'A Bourdieusian Analysis of Class and Migration: Habitus and the Individualizing Process'. *Sociology*, 44: 49–66.

Orange, C. 1987. *The Treaty of Waitangi*. Wellington: Allen & Unwin.

O'Reilly, K. 2005. *Ethnographic Methods*. 2nd edn. London: Routledge.

O'Reilly, K. 2000. *The British on the Costa del Sol: Transnational Identities and Local Communities*. London: Routledge.

Paterson, L. 2012. 'Pākehā or English? Māori Understandings of Englishness in the Colonial Period'. In *Far From Home: The English in New Zealand* edited by L. Fraser and A. McCarthy, 123–144. Dunedin: Otago University Press.

Pearson, D. 2000. 'The Ties That Unwind: Civic and Ethnic Imaginings in New Zealand'. *Nations and Nationalism*, 6: 91–110.

Pearson, D. 2008. 'Reframing Majoritarian National Identities Within an Antipodean Perspective'. *Thesis Eleven*, 95: 48–57.

Pearson, D. 2012. 'Arcadia Reinvented? Recounting the Recent Sentiments and Experiences of English Migrants to New Zealand'. In *Far from Home: The English in New Zealand* edited by L. Fraser and A. McCarthy, 145–164. Otago: Otago University Press.

Pearson, D. 2013. 'Comparing Cultures of Decline? Class Perceptions among English Migrants in New Zealand'. *New Zealand Sociology*, 28: 81–101.

Pearson, D. 2014. 'Ambiguous Immigrants? Examining the Changing Status of the English in New Zealand'. *Nations and Nationalism*, 20: 503–522.

Pearson, D. and Sedgwick, C. 2010. 'The English Question in New Zealand: Exploring National Attachments and Detachments among English Migrants'. *Nationalism and Ethnic Politics*, 16: 443–464.

Phillips, J. and Hearn, T. 2008. *Settlers: New Zealand Immigrants from England, Ireland and Scotland,1800–1945*. Auckland: Auckland University Press.

Pocock, J. 1975. 'British History: A Plea for a New Subject'. *The Journal of Modern History*, 47: 601–621.

Samuel, R. and Light, A. 1998. *Island Stories: Unravelling Britain, Theatres of Memory*. London: Verso.

Schech, S. and Haggis, J. 2004. 'Terrains of Migrancy and Whiteness: How British Migrants Locate Themselves in Australia'. In *Whitening Race: Essays in Social and Cultural Criticism* edited by A. Moreton-Robinson, 176–207. Canberra: Aboriginal Studies Press.

Spoonley, P. 1991. 'Pākehā Ethnicity: A Response to Māori Sovereignty'. In *Nga Take: Ethnic Relations and Racism in Aotearoa/New Zealand*, edited by P. Spoonley, D. Pearson and C. Macpherson, 154–170. Palmerston North: Dunmore Press.

Spoonley, P. and Bedford, R. 2012. *Welcome to Our World? Immigration and the Reshaping of New Zealand*. Auckland: Dunmore Pub.

Statistics New Zealand. 2013a. '2013 Census DataHub on Demand: Birthplace'. *Statistics New Zealand*. [Online] Available at: http://nzdotstat.stats.govt.nz/wbos/Index.aspx?DataSetCode=TABLECODE8048# [Accessed 19 February 2016].

Statistics New Zealand. 2013b. '2013 Census QuickStats about Culture and Identity: Ethnic Groups in New Zealand'. *Statistics New Zealand*. [Online] Available at: www.stats.govt.nz/Census/2013-census/profile-and-summary-reports/quickstats-culture-identity/ethnic-groups-NZ.aspx# [Accessed 12 February 2016].

Ward, P. 2004. *Britishness Since 1870*. London: Routledge.

Weight, R. 2002. *Patriots: National Identity in Britain, 1940–2000*. London: Macmillan.

Wellings, B. 2011. 'The English in Australia: A Non-Nation in Search of An Ethnicity?'. In *Britishness, Identity and Citizenship: The View from Abroad* edited by C. McGlynn, A. Mycock, and A. McAuley, 249–266. Oxford: Peter Lang.

Williams, C. 1995. '"Race" and Racism: Some Reflections on the Welsh Context'. *Contemporary Wales: An Annual of Economic and Social Research*, 8: 113–131.

Williams, C. 2005. 'Problematising Wales: An Exploration in Histiography and Post-coloniality'. In *Postcolonial Wales* edited by J. Arron and C. Williams, 3–22. Cardiff: University of Wales.

Wills, S. 2005. 'Passengers of Memory: Constructions of British Immigrants in Post-Imperial Australia'. *Australian Journal of Politics and History*, 51: 94–107.

Wills, S. and Darian-Smith, K. 2003. 'Beauty Contests for British Bulldogs? Negotiating (Trans)National Identities in Suburban Melbourne'. *Cultural Studies Review*, 9: 65–83.

Young, A. 2004. 'Experiences in ethnographic interviewing about race: the inside and outside of it'. In *Researching Race and Racism* edited by M. Bulmer and J. Solomos, 187–202. London: Routledge.

9 Size matters

British women's embodied experiences of size in Singapore

Jenny Lloyd

Introduction

> [W]hile migration studies may have long been about the movement of bodies, it cannot be said that bodies have been a prominent spatial scale of analysis in the field.
>
> (Dunn 2010, 1)

Work on migration has, to a large extent, been disembodied. Indeed, the prominence of economic perspectives has meant that migration has often been explored through a macro-level lens. This work, as Dunn (2010) suggests, has failed to account for the everyday practices of migrants. Furthermore, research that has focused on privileged migration has often painted migrants as hyper-mobile agents devoid of socially and culturally inscribed embodied markers and has ignored the relations between different people in place (Sklair 2000). Recent research however, has highlighted the different ways that everyday practices and spaces of migration are inherently embodied (Conradson and Latham 2005, Ho and Hatfield 2011). More broadly, Dunn (2010) has argued that embodied approaches account for the varying degrees of access afforded to migrants across different identity markers. Indeed, interest in embodied subjectivity has exemplified how gendered and racialised processes are lived and experienced (Silvey 2004). For example, research has explored everyday encounters within global cities and how these are understood through a gendered lens (Yeoh and Willis 2005). Work on British migration has noted the varying ways that gender, ethnicity and nationality intersect, contributing to performances of nationality that both reconstruct and challenge British imperialism (Leonard 2008).

It is not only within work on migration that the experiences of bodies, and their materiality, have been marginalised. Indeed, body size, has failed to gain significant geographic scrutiny beyond mapping 'obesity epidemics' (Boero 2007). Here, research has often been pre-occupied with theorising bodies through quantitative approaches that do not capture the multiplicity of sized experiences. Yet, there has been significant attention given to medical and popular discourses surrounding bodies of different sizes. Indeed, many societies within the 'West' today are in the grip of concerns regarding the *obesity epidemic*. In the UK, there is a proliferation of popular and medical 'knowledge' that suggests that fatness is not only undesirable and unattractive, but unhealthy and dangerous. For many people these discourses have significant implications to how they experience their own bodies. However, it has not been until recently that academic work has begun to interrogate these experiences from beyond a medicalised approach (Longhurst 2005). To explore how it *feels* rather than what it *means* to be sized. In this chapter I explore what happens when these two areas are brought together – migration and body size – to provide a broader framework in which to explore British migration and the multiple ways that body size is shaped and experienced.

This chapter responds to calls for greater work on embodied migration and work within Fat Studies that highlights the need for greater cross-cultural work on size (Cooper 2009). I explore transnational migration and body size by focusing specifically on British women expatriates in Singapore. By looking at one particular aspect of embodiment – body size – this chapter brings to light the multiple and intersecting relations that surround it. While there are multiple ways to conceptualise British migration, embodiment provides a framework through which to explore how people live their lives through their bodies and recognise, not only the place of structural forces within migration, but the multiple ways subjects are made through movements (Silvey 2004). In this chapter I foreground the experiences of British women by placing bodies as central to their experiences and everyday perceptions. I do this by *transsizing* – exploring the intersections between migration and experiences of body size (Lloyd 2014).

This research focuses on a specific group of privileged migrants – British expatriate women. While they have received less academic scrutiny than those from developing countries, the experiences of this group are important to how we theorise both British migration and that from a gendered perspective (Willis et al. 2002). As Fechter (2010) argues, by only focusing on one group of migrants we risk perpetuating a limited view of global migration. As such, the position and experiences of those defined as privileged migrants often result in homogenous assumptions regarding their positioning within the world (Fechter and Walsh 2010). Throughout this chapter I use the term 'expatriate' to refer to a particular group of privileged migrants in Singapore. Following from Cranston (2017) I recognise that 'expatriate can be seen as a way through which British migrants categorise and order migration, producing themselves as being "good migrants" both in the context of immigration debates in the UK and a growing hostility towards immigration in Singapore' (p. 2). Within the context of Singapore it is clear that the term expatriate has postcolonial connotations and is often used to situate people within a particular community in opposition to other migrant groups. While not unproblematic, I use this term throughout as my participants have done, to define a group of privileged migrants within Singapore. I do so to recognise both how multiple sites of privilege – ethnicity, class, economic capital – may intersect within migrant experiences of Singapore, but also to highlight the multiplicity of 'expatriate' experiences, challenging the homogenous stereotypes that pervade how many privileged women are discussed in popular discourse.

This chapter provides new ways of understanding postcolonial encounters within the context of discussions of size by arguing that body size discourses provide a language and means through which people can situate difference and their emotional experiences of migration. Specifically, I address the everydayness of size by focusing on some of the practices and narratives within which British women's sized identities are constituted, and the implications this has for how they discuss migration. The first section of the chapter introduces work on body size and migration. I then turn to the empirical analysis where I focus on two aspects of body size; firstly clothes sizing and quantification of bodies, and secondly relational sized encounters and racial discourse.

Trans-sizing – body size and migration

While there is growing work on embodied transnationalism, far fewer studies have explored the significance of body size. Current attention to embodied subjectivities and work on body size is theoretically underpinned by earlier work by feminists seeking to bring bodies into research (Longhurst and Johnston 2014). This work argued that the Cartesian separation of mind and body has led to fatness being treated as a facet of the unruly and devious body that should be disciplined (Evans 2006). Within the 'West', much of the rhetoric that surrounds fatness is dominated by medical understandings of fatness as bad, unhealthy and unattractive (Braziel and LeBesco 2001). However, there is work that explores fatness and body size through a critical lens, bringing a range of voices and experiences into discussions of size (Rothblum and Solovay 2009). Work within this by social and cultural geographers, looks to expand academic and popular assumptions about fatness and body size by listening to the 'often private, sometimes silenced and somewhat concealed experiences of people with different body sizes' (Hopkins 2008, 2112). In order to explore British women's experiences of migration I grounded this research within feminist theoretical work on embodiment. Embodiment has enabled feminist academics to deconstruct masculinist binaries of knowledge production (Rose 1993), and make room for understanding the ways that gender is lived in and through material and discursive bodies (Longhurst and Johnston 2014).

However, so far few studies have critically explored the significance of transnational migration to experiences of body size (Cooper 2009). Utilising cross-cultural perspectives would enable academics to expand understandings of body size and how people make sense of and experience it every day. In response to this gap, I utilise the concept of trans-sizing to explore how cross-cultural perspectives are important to the subversion or reproduction of dominant ideas regarding fatness. Furthermore, it provides an important lens in which to examine the multiple ways that fatness intersects with gender, ethnicity, nationality, class, sexuality, culture and religion. In this chapter I highlight the ways that body size is bound to racialised and classed constructions of 'legitimate' bodies by focusing on expatriate British women's experiences of migration. Before exploring the empirical findings, I outline the research context.

British expatriates in Singapore

Singapore is a relatively young country. It became an independent republic in 1965 after receiving self-governing status from Britain in 1959 following 140 years of British colonial rule (Perry, Kong, and Yeoh 1997). From its establishment as a British trading post by Stanford Raffles in 1819, colonialism and migration have shaped, and continue to shape, much of Singapore's contemporary form today. As of 2015, 29% of the population of Singapore were recognised as non-resident (Singstat. gov 2015), and in 2013 the UK Government estimated that there were 30,000 British nationals living in Singapore. As a 'postcolonial global city', local social, cultural, economic and political processes intersect with transnational movements, flows and migrations (Yeoh and Chang 2001). Yet, while the notion of a global city itself is imbued with meaning regarding mobility and movement, it has been noted that this is at times in conflict with the nation-building agenda that Singapore must negotiate (Ho 2006). As a postcolonial city, Singapore provides an important context in which to explore migrant relations and the interactions between people in place. In this chapter, I utilise a postcolonial perspective to highlight the ways in which colonial relations are reconfigured, mobilised and narrated by expatriates. Following from Fechter and Walsh (2010, 1198) to argue that 'such a perspective might reveal how racial hierarchies and power inequalities persist, as well as how they are being reconfigured and challenged'.

Critical academics have highlighted that cities are not only sites of economic capital but shaped through gendered processes (Yeoh, Huang and Willis 2000). There is a body of work that has researched migrant relations within Singapore, including studies exploring this from a gendered perspective (Yeoh, Huang and Devasahayam 2004, Kitiarsa 2008, Huang and Yeoh 2003). While this important work provides a framework for recognising women's experiences of migration within Singapore, it is fair to say that there has been limited attention, both in work on Singapore and elsewhere, on privileged women's experiences of migration (Willis, Yeoh and Fakhri 2002). I argue that in order to critique and engage with postcolonial narratives of Singapore it is essential that we account for a range of voices, contributing to a rich and developed discussion of different people's lives within the postcolonial city (Kunz 2016). In the following section I provide a summary of the research methodology.

Methodology

The data used in this chapter is based on empirical qualitative research carried out within a six-month period of fieldwork in Singapore between 2012 and 2013. The project involved research with women who self-defined as expatriates, 50% of whom were British and whose experiences form the basis of this chapter. In order to *get at* these experiences of body size, in-depth interviews were carried out with women about their experiences of migration. This chapter is based on 25 in-depth interviews and one focus group with British expatriates. The age of the participants interviewed varied from 30 to 60 years, with a wide-ranging length of expatriation to Singapore from 6 months to 18 years. Twenty-three of the participants defined their ethnicity as white, Caucasian or English with one participant identifying as of Irish ethnic heritage and one as British-Indian. All but two of those interviewed migrated due to their partner's jobs. The methods used within this project were shaped by the intersections of feminist methodological approaches and work within Fat Studies. This meant that rather than seeking to quantify size I have utilised qualitative approaches to explore the multiple and conflicting ways that body size is experienced at different times and in different places. By doing so I focus on the subjective experiences of women by valuing the ways women construct their own biographies and tell stories about their lives. I suggest that methodologies that value, and are representative of, nuanced understandings of migration and body size provide a progressive means through which to critique the ways that people experience their social worlds through their (sized) bodies (Lloyd and Hopkins 2015). I move now to the empirical findings of this research.

Body size and belonging

At the start of this chapter I suggested that in the UK there is a dominance of anti-fat rhetoric surrounding bodies, what Murray (2005, 154) suggests is a 'culture of a negative collective "knowingness" about fatness'. The same can also be said for Singapore, albeit manifesting in different ways (Isono, Watkins and Lian 2009). By being situated within transnational spaces women continue to be informed by multiple discourses of size both within Singapore 'here' and the UK 'there'. These discourses are not equal; women's bodies are particularly vulnerable to the disciplinary practices and surveillance of obesity rhetoric as medical discourses often unite gendered ideas regarding health, with historical and patriarchal understandings of women's roles (Giovanelli and Ostertag 2009). Within the context of migration, feminist academics have urged scholars to take note of how practices and spaces of migration contribute, reinforce and deepen marginalisation, particularly women's experiences (McEwan 2004). Within this research many women stated that migration impacted their sense of identity, suggesting that this was the result of having to leave their jobs for their partner's career. The following comment by Susan is typical of reflections that many women made:

You lose, yeah you lose yourself, [...]. You lose everything you were, because I guess, we all know who we are in our own little context [...]. And then to come here I just felt like that was all stripped away from me. [...] You're starting from the ground zero.

(Susan, 40, freelance writer, married[1])

While Susan is reflecting here on many aspects of migration including work, friends and her social life, she also suggested that moving to Singapore, and losing her sense of self, had impacted how she managed her feelings towards her size. Although it can be argued that there is extreme pressure on women to look a certain way in the UK, for many of the women I spoke with – like Susan – transnational migration marked a significant change in their sense of identity through the production of new subjectivities and the ways they experienced their size. Most explicitly, this was often discussed in relation to the perceived size of 'Asian' women who were considered smaller and slimmer. I argue that the mechanisms that women may utilise to cope with body size discourses are altered under conditions of migration, often through a feeling of loss of identity but also through relational interactions and experiences. In the following extract, Beth discusses this further and several of the ways that she experiences her size as a British expatriate in Singapore:

And the other side of body image – whiteness I said first, but there's also, just, being a big clumpy Western woman. In a world which seems to be populated by tiny, petite, beautiful, perfectly groomed people. [In Singapore] you're different, you *know* you're different, you're being measured by a different set of standards and however much you say people aren't thinking about it [appearances] – actually they are. [...] in the UK it's the same, people *are* commenting on people that walk by. And in the UK I just don't give a toss, I just shrug it off, because you know, because at the end of the day I know who I am there. But I don't here.

(Beth, 46, dependent, married)

For Beth, it is clear that migration has not only impacted how she feels about herself but also her ability to cope with and resist discourses regarding body size that normally she could 'shrug off' in the UK. She suggests that her inability to do so is due to not knowing who she is in Singapore. While it is not clear if Beth is being judged differently, and later she states, 'it could just be in my mind' in reference to her whiteness and 'sticking out', it is clear that body size discourses are embodied differently within this new context. Within this transnational space UK standards still inform ideas about size but simultaneously intersect with complex, contradictory and shifting ideas about women's identities through migration. In what started out as a discussion of body size, has opened up avenues in which Beth is able to disclose personal narrations of her experiences of migration. While many factors are at play, many others linked this sense of loss of identity to not being able to work in Singapore and having to rely financially on their husbands:

> I can't do anything without my husband. [...] You're not really worth anything because you're not earning any money because technically you're a dependant on somebody else to, for financial stability.
>
> (Anita, age unknown, unemployed, married)

For the women I spoke with, body size and fatness materialised as 'central to the subjective experience' of migration (Colls 2007, 359). One way that this was explicitly highlighted was through discussions of clothes shopping and sizing. In fact, talking about the often-mundane aspects of body size and clothes shopping provided particularly revealing and interesting narratives regarding difference that may otherwise not have been discussed. Talking about clothes shopping provided an accessible way through which to talk about body size without being intrusive or insensitive. In the following section I explore how clothes shopping, and specifically clothes sizing, are ways in which women's bodies are materially and discursively marked as outsiders within Singapore. I focus firstly on clothes sizing specifically and then discuss relational size and racial discourses.

The 'here and there' of sizing

It was well established within the expatriate communities I engaged with that large disparities exist between clothes sizes in Singapore, even when multiple international sizes are displayed on the label. Therefore many, if not all, of the women found that despite displaying the UK size, this had little correlation to the size they would wear in the UK (for the most part it was often smaller). For many of the women, this seemingly small detail was often the source of frustration during and after clothes shopping when women were unable to find clothes that fit, being 'forced' to buy a larger size. The implications of this appeared to shape not only their feelings towards their bodies at the time but also how they understood and constructed their new femininities as expatriate women (Walsh 2007), and how they imagined and constructed ideas about what constitutes a Singaporean body:

JENNY: Do you ever clothes shop here?
MARTHA: No, it's too depressing, it's too expensive and everyone's tiny.
(Martha, 51, administration manager, divorced)

Sharon suggests that although clothes may be categorised using European or UK sizing, it was often assumed that the cut of the clothes is such that it suits an 'Asian' physique rather than that of an expatriate (a white European or American expatriate):

So when you get a size 12 or a 14 or whatever you wear and you put it on, it might fit you but then you can't drive because the arm holes are in the wrong place or the jacket, the waist is in the wrong place, because it's a completely different cut because they cut it for the Asian size.

(Sharon, 49, designer, married)

The quantification of body size was not only important to find clothes that fit and were comfortable, but can be seen to have contributed to the embodiment of size through this quantification and the imagined idea of what an 'Asian' body is. Thus body size materialises through sizing practices and is subsequently imagined, felt and experienced as a result of conflicting and overlapping global networks of normalised understandings of what is an acceptable body size. Colls (2004, 588), suggests that '[N]umbers or numerical sizing provide the means for women to solidify their body's material form and fix it both spatially and temporally'. For many of the women I spoke with, numerical sizing played a significant role in their embodied experience of size when migrating. For many, the clothes size they were able to physically fit into dictated the type of emotional response they had while clothes shopping, as it 'enables women to make links between past and present, emotional as well as physical, "well-being"' (ibid.). For many of the women, disparities between clothing sizes in Singapore and the UK, contributed to them feeling negatively about themselves and their body size. Where Colls (2004) proposes sizing is a way in which to link 'between past and present' and 'spatially and temporally', it is my suggestion that within the context of transnational migration, linkages are also made between home and host, embodying a particular national understanding of what size is and what size the women 'should' be. For many, there was a need to 'fit' within clothes in the size they would wear in the UK. I argue that this idea of *fitting* is not only about the practicalities of finding clothes to wear but highlights the extreme pressure on women to *fit* within sizes that are recognised as socially acceptable in the UK. Thus not fitting appears to mark women's bodies both outside of what is considered normative sizing in the UK and in comparison to other bodies in this new context.

The lack of continuity between sizing in Singapore and the UK often contributed to the women experiencing size negatively. Many of the women I spoke with talked about these experiences by discussing and reminiscing about the size they were in the UK and the size they now wear in Singapore:

JENNY: Do you think that you when you first moved here or now, that it changed the way that you think about yourself?

SUSAN: Yeah, completely, I mean I've always been like a size 12 or 14 [in the UK], and I've always worked out, it's just one of those things that you are the frame you are [sighs]. Then when I got here, you just feel like a giant, you just go in and you're looking at XXXL sizes, thinking *God*.

(Susan, 40, journalist, married)

It is clear that the numerical sizing used in Singapore impacts the women's experiences of shopping and ultimately how they feel about their bodies. While the women understood that sizes were different from the UK this did not seem to mitigate them experiencing this negatively. In fact, for many there was a clear correlation between the numerical size and feeling negatively about themselves. For many, X (extra) sized clothing had negative connotations in the UK and as such contributed to labelling their bodies as '"outside" of normative notions of style and sizing' in Singapore (Colls 2006, 537). It is clear from Susan's comment that despite accepting her body (and her height) in her home country, like Beth, migration has cast her body outside of what she considers normative sizing. What is important is the affect that this quantification has to how she feels. Despite not becoming physically fatter or taller, she embodies her corporeality in a way that makes her feel bigger – 'like a giant' – as her size and stature are materialised through sizing her body.

Clothes sizing serves as a way through which people are able to monitor and regulate their body size, so as to fit within normative constructions of what constitutes a 'legitimate' and 'healthy' body within society (Evans and Colls 2009). Measures of numerical sizing therefore serve as a bio-political strategy that enables women to monitor their body through sizing, and deem whether it is acceptable or not through the knowledge they have gained from dominant popular and medical discourses regarding fatness (Wright 2009). Despite knowing that sizing is different to that in their home countries, the pervasiveness of normative ideas of what is an acceptable body size, is such that it is able to overrule the 'objective' understanding that 'you're not big … you're in Asia' (Sharon), and induce feelings that are 'depressing' and 'like a giant'.

For many of the women I spoke with, clothes sizing was not just about purchasing clothes that fit, but about the *here and there* of sizing. National identity was reinforced through the use of numerical sizing as a point of reference in order to quantify body size in relation to the sizing used at home. For many, clothes sizing contributed to a banal nationalism where numerical sizing acted as a point of reference for which nationality (and difference) were experienced, remembered and practised (Billig 1995). That is to say, that while often unnoticed within our home context, activities – in this case dressing and clothes shopping – become routine, where fitting into certain clothes sizes may become habitual. As such, these processes and the symbols related to them, become associated with past experiences and embodied processes of 'fitting' – both physically in terms of the materiality of bodies and socially in terms of the different sizes. Clothes sizes may be particularly imbued with discourses of national identity because they often differ in different places (i.e. UK, European, American) or may remain as measurement of size following colonial rule. For many of the women I spoke with, size was quantified through the use of numerical sizing used in their home country despite living in Singapore for years. This type of embodied knowledge was often necessary when buying clothes back home. Furthermore, it appeared to become both a point of solidarity for many expatriate women when discussing some of the difficulties of adjusting to moving to Singapore, and part of the boundary-making process between self and other (Hindman 2009). In my own experience of attending expatriate social events, many of the first things women would talk about were their problems buying clothes, advice on shops to go to, and the size of Singaporean women. In the following section I focus on this latter point.

Relational size and racial discourse

In addition to the women's own experiences of clothes sizing, perceptions of size were impacted by relational encounters with other bodies. In the following extracts I highlight how anxieties about size and 'local' women are the product of intersections of different forms of discrimination and subordination that have solidified representations of some women as exotic *and* erotic others (Teo and Leong 2006). To a certain degree, the anxieties and fears expatriate women may have regarding other Asian women can be seen to embody and construct Otherness through 'hegemonic masculine fears and fantasies about the feminine Other' (Hubbard 1998, 57). In the postcolonial context, it is not just a matter of difference, but *Otherness*. Colonial racial and sexual discourses construct and fix difference as otherness through the repetition of stereotypical discourses (Bhabha 1996). In particular, through the pervasiveness of stereotypes regarding Asian women as 'hyperfeminine: passive, weak, quiet, excessively submissive, slavishly dutiful, sexually exotic, and available for white men' (Pyke and Johnson 2003, 36). These sexual and racial discourses have been used to justify the separation and opposition of the Western female self, from the oriental Other. Accordingly, unequal power structures polarise identities and reinforce binary relations between women, essentialising the dichotomy of Western/Asian (Bulbeck 1998). However, as Hubbard (1998) argues, othering is not only through discursive constructions of difference, but is carefully produced through the visual as well. Many of the anxieties that arise from these (assumed) differences are a product of the intersections of dominant colonial discourses and the visual consumption of bodies and their performances of sexuality. As Kaplan (1997) argues, this places white women in a contradictory position where they are simultaneously and paradoxically both the subject of white male gaze and colluding with it, through their objectification and discrimination of Asian women. I argue that, anxiety is also performed and narrated through the embodiment of the white male gaze, and dominant discourses that discipline women's bodies to be thin in order to be considered sexually attractive. As such, I suggest that the women narrate their anxiety in multiple ways, for example through the positioning of their 'fat' bodies in opposition to the 'thin' bodies of Asian women.

During interviews, discussions of size often alluded to racialised practices of othering, through talking about their bodies in relation to other (non-white) bodies. The following extract suggests the ways that women embodied their size relationally and the processes of othering produced through this. This extract is taken from a focus group I conducted with members of a weight-loss organisation aimed at UK expatriates:

JENNY: And I just wondered why you had joined [group name]?

PAT: The locals are so slim, it just hits you, you don't notice it in the UK. You just see them here, they're stick thin.

GAIL: You go into a shop and you have to get an XXL it's just a bit depressing.

ELIZABETH: I just felt that when I got here that my husband would never look at me again. Because, what with all these gorgeous girls why would he look at me? I mean what was there to look at? I was the fat, I felt so frumpy and fat. Oh God.

JENNY: I think it's like, yeah, and also people dress up a lot more here.

PAT: Well they dress up to go to work, don't they? They think they're going to a disco. Oh, you don't have discos anymore – nightclub.

EVERYONE: [Laughing]

PAT: They go to work in things like they're going to a nightclub or a wedding.

GAIL: Lacy things.

ELIZABETH: And you can't buy clothes can you because it's all designed for them not for us?

EVERYONE: [Assent]

ELIZABETH: The taste is completely different.

PAT: It's terrible.

> (Pat, 54, housewife, married; Gail, 50, married; Elizabeth, 60, retired, married)

Global cities like Singapore provide multiple opportunities for people to come into contact with one another. Within these *contact zones* difference becomes proximate, and otherness materialises through social, temporal and spatial, relations and interactions (Yeoh and Willis 2005). These encounters are important because they demonstrate how identities are formed through everyday encounters with others (Leonard 2010a). In this extract, it is clear that relational encounters within Singapore shape how the women experience themselves. Firstly, the women wish to lose weight as a direct result of feeling fatter than 'the locals'. Second, they position 'local' women in opposition to them through sexualised narratives and stereotypes. Finally, discussions regarding their own experiences of size are narrated through racialised discourses of (non-white) local women as Other (Leonard 2010b). Again, this extract highlights how discussions of size open up opportunities for discussing experiences of migration in different ways.

The motive many of the women had for joining the group and wanting to lose weight is discussed as a direct result of moving to Singapore and the size of the 'locals'. Although they never explicitly state the nationality of the people they are referring to, it is inferred throughout the discussion that they mean women of Chinese descent. Again, local does not necessarily indicate Singaporean nationality, rather an imagined idea of Asian women (Fechter 2016). This discussion highlights that not only do the women find it frustrating that the clothes are too small for them, but that this contributes to an imagined understanding of the size of local women. It is clear that encounters with local women within the contact zones of the city have contributed to their understanding of them as 'stick thin' and 'slim' but also having 'terrible taste'. While this is about seeing and viewing bodies, it is also about imagining them and constructing ideas based on established racialised and classed discourses. Within Singapore, shops are contact zones within which women come into the contact with others, but also imagine the presence of others through the use of sizing.

All of the women in the focus group and almost every woman I spoke with reflected on the size of local women using a variety of adjectives. Here Pat suggests that the women are 'stick thin', but many of the other women that I spoke with referred to Asian women through a range of nicknames they had invented in relation to their size, for example: 'Micro-Asian ladies', 'Stick insects', and 'Flat-packs'. While it is clear that many of the women seek to be thin themselves, I would argue that through the use of these terms they position 'local' women paradoxically as both subjects of envy, and simultaneously subsume racial and gendered stereotypes that represent them as inferior and the subject of jokes. Furthermore, the women also appear to reclaim their sense of superiority through discussions of style and class by deeming Singaporean taste as 'terrible' and the suggestion that women dress as though they are going to a 'nightclub', which can be seen to invoke particularly gendered and sexist ideas around women's clothes, sexuality and class. The use of these terms clearly represent the processes of othering which identify 'local' or 'Asian' women as passive others. Additionally, I suggest that through this they are, to some extent, able to emotionally manage their relational experiences of body size. Clearly there is a large impetus for women to lose weight within the UK. However, it is clear that migration to Singapore and relational encounters with people of different ethnic origins intensifies the way that difference is marked on the body and played out through daily narrations and performances.

It is clear that these discussions were common for expatriates in Singapore. These terms and observations highlight the dominance of globalising discourses of the thin ideal, but also the subtleties and disciplinary regimes that women must also not be *too* thin. Following from Fechter (2010), I suggest these comments hint at the 'sexual jealousy of "foreign" women' (p. 1288) which may arise out of the proximities of different body sizes but are also a result of colonial processes aimed at promoting the separation of different racial groups (see Fechter 2010). This is not simply a result of expatriate women wishing to be thin. Instead, these terms highlight how relational encounters both shape how women see themselves and also how persistent sexist and racial stereotypes perpetuate ideas regarding Asian women as sexual predators within the postcolonial global city. These insecurities are highlighted in Elizabeth's previous comments regarding her husband's gaze. While Elizabeth does not overtly suggest that local women are sexual predators, she discusses her implicit assumption that her husband will no longer find her attractive. As argued earlier, for many women, migration destabilises their sense of self-worth. The comment by Elizabeth can be seen to augment the understanding that a woman's primary role is as a sexual object, and can be seen as participating in a dominant expatriate narrative regarding husbands' infidelity (Yeoh and Huang 2010).

It is clear through the references to women as 'girls' and the comment about not having discos any more (which felt directed at the researcher as well as younger women in general) that age is a key factor in how the women experience their new subjectivities within Singapore. While some comments may have been made out of an awareness of the age difference between the participants and myself, it was clear that many of these reflected the intersection of age, gender and whiteness. When Elizabeth states 'with all these gorgeous girls why would he look at *me*?' she is referring specifically to 'local' women. There is an insinuation here that Singaporean women are a concern because it is assumed that they are young, single and both in pursuit of white men and sexually available. Thus, feeling 'frumpy and fat' can be recognised as a product of relational interactions and the intersections of class, age, ethnicity, gender, sexuality and body size within postcolonial relations that have historically, and continue to, position expatriates and local women in opposition. At the same time postcolonialism underscores the fluidity of power within colonial relations – that in many cases the powerful are simultaneously powerless. So feeling frumpy and fat reveals how powerless these women feel in the new context and within patriarchal relations. Arguably, 'name calling' the local women and judging their sense of style could be seen an act of resistance.

How the women discussed, related to and imagined Singaporean women was often through the use of stereotypes and visual markers of difference. Take for example the following quotes from Sharon and Sophie:

> It's very difficult for the women because you go out to an office party or whatever and there are all these cute little girls that look 15, they're actually much older. But they look so cute and they look so young and they're so tiny and you go there, you've just had three babies and you feel like the stuck at home *mother*. (Sharon, 49, designer, married)
>
> They're all very small and very slim. They all take a lot of care of themselves. They're all into the shopping and their fashion and their style. I mean some of the women their outfits are fantastic. Everyone looks like a model, and it's like *oh my God* and it's like some days I'm just like I don't feel like dressing up, some days I feel like a giant on the MRT [Mass Rapid Transit].
>
> (Sophie, 35, housewife, married)

I argue that body size is a significant way through which racialised and gendered discourses of difference are played out. Thus difference is discussed, not overtly through racial narratives, but through 'legitimate' conversations regarding the body size of Singaporean women. Relational encounters cement the understanding that expatriate and local women are different. For example, that they have different tastes – that are 'terrible', and different bodies, 'you can't buy clothes can you because it's all designed for them not for us'. Body size acts as another way through which difference is embodied and asserted between different groups within Singapore; a result of both contemporary Western discourses regarding thinness, and historical colonial relations regarding expatriates and local women.

This is not to say that anxieties exist for all women, but it does highlight some of the structural processes that shape women's relationships with one another, and some of the dominant discourses that pervade expatriate discussions regarding local women. Fechter's (2010) work highlights the ways that contemporary tropes regarding white expatriate women as concerned about, and threatened by Asian women, can be traced back to colonial discourses. I would argue that in these comments about 'Asian' women, there is clear resonance between colonial discourses and contemporary versions of how different groups of women are viewed, which are maintained through gendered, racialised, classed and sexualised othering practices within the postcolonial city. Additionally, as the comments above highlight, the materiality of bodies is also important. Indeed the size, shape and proximities of different bodies can shape the ways through which difference is experienced and embodied.

Conclusion

Throughout this chapter I have explored British migration through a unique lens – body size. I have highlighted new ways to understand postcolonial encounters by exploring how British women situate difference and emotional experiences of migration through body size narratives. While multiple and intersecting factors are at work in how British women experience migration, body size and fatness play a key role in their everyday narrations and embodied experiences of Singapore, allowing them to articulate a range of migrant experiences. This highlights the significance of women's experiences, the role of body size within this, the power of normalising discourses around body size and exemplifies the diverse stories and narratives that are produced when scholars take note of and value size.

Women's experiences, and often 'privileged' women, have largely been absent from academic work on migration (Walsh 2007). The same can be said for body size. What I have shown throughout this chapter is the importance of recognising and listening to experiences that are often overlooked. Ultimately, discussions of size opened up the opportunity to talk about British experiences of migration and difference in distinctive ways. For example, talking about shopping and the quantification of bodies through sizing, contributed to negative experiences of the women's size *and* Singapore, while further deepening their sense of being outsiders. Clothes sizing draw upon national symbols that augment dominant discourses around bodies in the UK that are particularly gendered. The pressure placed on women to be thin in the UK shapes, and continues to shape, how women experience their bodies in Singapore. Yet, in the context of migration, construct difference across and between women of different races. Encounters of difference (both real and imagined) within the global city are central to how people embody their own sense of identity. I have shown how discussions of body size are a means through which to explore how these encounters were talked about and how these intersected with other discussions of difference. Critically, body size discussions are utilised as a 'legitimate' means through which to do so. Thus, racialised and sexualised narratives of difference are subsumed within body size discourses that position Asian women as subservient and in opposition to British women. While body size is clearly not the only way to explore British migration it is clear that these experiences are central to some migrant narratives in constructing and reinforcing postcolonial notions of difference. Trans-sizing has opened up the possibilities for exploring issues that may otherwise be overlooked. This presents new opportunities for work that interrogates experiences of migration.

Note

1 The demographic information is presented as my participants defined themselves and includes their age, occupation in Singapore and marital status. All names have been changed.

References

Bhabha, H.K. 1996. "The other question: difference, discrimination, and the discourse of colonialism." In *Black British cultural studies: a reader*, edited by Manthia Diawara, Houston A. Baker and Ruth H. Lindeborg, 87–106. Chicago: Chicago University Press.

Billig, Michael. 1995. *Banal nationalism*. London: Sage.

Boero, N.C. 2007. "All the news that's fat to print: the American 'obesity epidemic'." *Qualitative Sociology*, 30(1): 41–61.

Braziel, Jana Evans and LeBesco, Kathleen. 2001. *Bodies out of bounds: fatness and transgression*. Berkeley: University of California Press.

Bulbeck, Chilla. 1998. *Re-orienting western feminisms: women's diversity in a postcolonial world*. Cambridge and New York: Cambridge University Press.

Colls, Rachel. 2004. "'Looking alright, feeling alright': emotions, sizing and the geographies of women's experiences of clothing consumption." *Social and Cultural Geography*, 5(4): 583–596. doi:10.1080/1464936042000317712.

Colls, Rachel. 2006. "Outsize/outside: bodily bignesses and the emotional experiences of British women shopping for clothes." *Gender, Place and Culture*, 13(5): 529–545. doi:10.1080/09663690600858945.

Colls, Rachel. 2007. "Materialising bodily matter: intra-action and the embodiment of 'fat'." *Geoforum*, 38(2): 353–365. doi:10.1016/j.geoforum.2006.09.004.

Conradson, David and Latham, Alan. 2005. "Transnational urbanism: attending to everyday practices and mobilities." *Journal of Ethnic and Migration Studies*, 31(2): 227–233. doi:10.1080/1369183042000339891.

Cooper, Charlotte. 2009. "Maybe it should be called fat American studies." In *The fat studies reader*, edited by Esther Rothblum and Sondra Solovay, 327–333. New York and London: New York University Press.

Cranston, Sophie. 2017. "Expatriate as a 'good' migrant: thinking through skilled international migrant categories." *Population, Space and Place*, 23(6).

Dunn, Kevin. 2010. "Embodied transnationalism: bodies in transnational spaces." *Population, Space and Place*, 16(1): 1–9. doi:10.1002/psp.593.

Evans, Bethan. 2006. "'Gluttony or sloth': critical geographies of bodies and morality in (anti)obesity policy." *Area*, 38(3): 259–267. doi:10.1111/j.1475–4762.2006.00692.x.

Evans, Bethan and Colls, Rachel. 2009. "Measuring fatness, governing bodies: the spatialities of the Body Mass Index (BMI) in anti-obesity politics." *Antipode*, 41(5): 1051–1083. doi:10.1111/j.1467–8330.2009.00706.x.

Fechter, A.-M. 2016. *Transnational lives: expatriates in Indonesia*. London: Routledge.

Fechter, A. and Walsh, K. 2010. "Introduction to special issue: examining "expatriate" continuities: postcolonial approaches to mobile professionals." *Journal of Ethnic and Migration Studies*, 36(8): 1197–1210.

Giovanelli, D. and Ostertag, S. 2009. "Controlling the body: media representations, body size and self-discipline." In *The fat studies reader*, edited by E. Rothblum and S. Solovay, 289–298. New York andLondon: New York University Press.

Hindman, Heather. 2009. "Shopping in the bazaar/bizarre shopping: culture and the accidental elitism of expatriates in Kathmandu, Nepal." *The Journal of Popular Culture*, 42(4): 663–679.

Ho, E.L.E. 2006. "Negotiating belonging and the perceptions of citizenship in a transnational world: Singapore, a cosmopolis?" *Social and Cultural Geography*, 7: 385–401.

Ho, Elaine Lynn-Ee and Hatfield, Madeleine E. 2011. "Migration and everyday matters: sociality and materiality." *Population, Space and Place*, 17(6): 707–713. doi:10.1002/psp.636.

Hopkins, Peter. 2008. "Critical geographies of body size." *Geography Compass*, 2(6): 2111–2126. doi:10.1111/j.1749–8198.2008.00174.x.

Huang, Shirlena, and Yeoh, Brenda S.A. 2003. "The difference gender makes: state policy and contract migrant workers in Singapore." *Asian and Pacific Migration Journal*, 12(1–2): 75–97. doi:10.1177/011719680301200104.

Hubbard, Phil. 1998. "Sexuality, immorality and the city: red-light districts and the marginalisation of female street prostitutes." *Gender, Place and Culture*, 5(1): 55–76. doi:10.1080/09663699825322.

Isono, Maho, Watkins, Patty Lou and Ee Lian, Lee. 2009. "Bon fatty girl: a qualitative exploration of weight bias in Singapore." In *The fat studies reader*, edited by Esther Rothblum and M. Wann, 127–138. New York andLondon: New York University Press.

Kaplan, E.Ann. 1997. *Looking for the other: feminism, film, and the imperial gaze*. New York and London: Routledge.

Kitiarsa, Pattana. 2008. "Thai migrants in Singapore: state, intimacy and desire." *Gender, Place and Culture*, 15(6): 595–610. doi:10.1080/09663690802518495.

Kunz, Sarah. 2016. "Privileged mobilities: locating the expatriate in migration scholarship." *Geography Compass*, 10(3): 89–101. doi:10.1111/gec3.12253.

Leonard, P. 2008. "Migrating identities: gender, whiteness and Britishness in postcolonial Hong Kong." *Gender, Place and Culture*, 15(1): 45–60.

Leonard, P. 2010a. *Expatriate identities in postcolonial organizations: working whiteness*. Farnham: Ashgate Publishing Ltd.

Leonard, Pauline. 2010b. "Organizing whiteness: gender, nationality and subjectivity in postcolonial Hong Kong." *Gender, Work and Organization*, 17(3): 340–358. doi:10.1111/j.1468–0432.2008.00407.x.

Lloyd, Jenny. 2014. "Bodies over borders: the sized body and geographies of transnationalism." *Gender, Place and Culture*, 21(1): 123–131. doi:10.1080/0966369X.2013.791253.

Lloyd, Jenny and Hopkins, Peter. 2015. "Using interviews to research body size: methodological and ethical considerations." *Area*, 47(3): 305–310. doi:10.1111/area.12199.

Longhurst, Robyn. 2005. "Fat bodies: developing geographical research agendas." *Progress in Human Geography*, 29(3): 347–359.

Longhurst, Robyn and Johnston, Lynda. 2014. "Bodies, gender, place and culture: 21 years on." *Gender, Place and Culture*, 21(3): 267–278. doi:10.1080/0966369X.2014.897220.

McEwan, C. 2004. "Transnationalism." In *A companion to cultural geography*, edited by J. Nuala, J. Duncan and R. Schein, 499–512. Hoboken: Wiley-Blackwell.

Murray, Samantha. 2005. "(Un/be)coming out? Rethinking fat politics." *Social Semiotics*, 15(2): 153–163. doi:10.1080/10350330500154667.

Perry, Martin, Kong, Lily and Yeoh, Brenda S.A. 1997. *Singapore: a developmental city state, world cities series*. Chichester and New York: Wiley.

Pyke, Karen D. and Johnson, Denise L. 2003. "Asian American women and racialized femininities: "doing" gender across cultural worlds." *Gender and Society*, 17(1): 33–53. doi:10.1177/0891243202238977.

Rose, G. 1993. *Feminism and geography. the limits of geographical knowledge*. Cambridge: Polity Press.

Rothblum, Esther and Solovay, Sondra. 2009. *The fat studies reader*. New York and London: New York University Press.

Silvey, Rachel. 2004. "Power, difference and mobility: feminist advances in migration studies." *Progress in Human Geography*, 28(4): 490–506.

Singstat.gov. 2015. *Latest data 2016* [cited 27/07/162015]. [Online] Available at: www.singstat.gov.sg/statistics/latest-data#14.

Sklair, L. 2000. *The transnational capitalist class*. London: Wiley-Blackwell.

Teo, Peggy and Leong, Sandra. 2006. "A postcolonial analysis of backpacking." *Annals of Tourism Research*, 33(1): 109–131. doi:10.1016/j.annals.2005.05.001.

Walsh, Katie. 2007. "'It got very debauched, very Dubai!': heterosexual intimacy amongst single British expatriates." *Social and Cultural Geography*, 8(4): 507–533. doi:10.1080/14649360701529774.

Willis, Katie, Yeoh, Brenda and Fakhri, S.M.A.K. 2002. "Transnational elites." *Geoforum*, 33(4): 505–507. doi:10.1016/S0016-7185(02)00038–00036.

Wright, Jan. 2009. "Biopower, biopedagogies and the obesity epidemic." In *Biopolitics and the 'obesity epidemic': governing bodies*, 1–14. New York: Routledge.

Yeoh, B.S.A. and Chang, T.C. 2001. "Globalising Singapore: debating transnational flows in the city." *Urban Studies*, 38: 1025–1044.

Yeoh, Brenda S.A. and Huang, Shirlena. 2010. "Sexualised politics of proximities among female transnational migrants in Singapore." *Population, Space and Place*, 16: 37–49.

Yeoh, Brenda S.A. and Willis, Katie. 2005. "Singaporean and British transmigrants in China and the cultural politics of 'contact zones'." *Journal of Ethnic and Migration Studies*, 31(2): 269–285. doi:10.1080/1369183042000339927.

Yeoh, B.S.A., Huang, S. and Willis, K. 2000. "Global cities, transnational flows and gender dimensions, the view from Singapore." *Tijdschrift Voor Economische en Social Geografie*, 91: 147–158.

Yeoh, Brenda S.A., Huang, Shirlena and Devasahayam, Theresa W. 2004. "Diasporic subjects in the nation: foreign domestic workers, the reach of the law and civil society in Singapore." *Asian Studies Review*, 28: 7–23.

10 Resilience and social support among ageing British migrants in Spain

Kelly Hall

Introduction

Around 168,000 British state pensioners live in Spain (Thurley, 2014) and there are now well-established communities of British people living on the Spanish Costas. Many people in these communities are now ageing in place and becoming increasingly dependent and vulnerable due to a severe decline in health, the need for care and bereavement (Hardill et al., 2005; Hall and Hardill, 2016). This chapter explores the resilience of older British people in Spain, focusing on those who are in need of care or other types of social (and/ or financial) support. It focuses on the ways in which they utilise social networks, support structures and other resources during times of crisis or adversity.

'Resilience' is defined as the capacity to withstand and rebound from crises or disruptive life challenges (Walsh, 2006) and is strongly associated with good social networks. During the fourth age, social and support networks are crucial resilience resources as they can provide help, advice, companionship and even validate an individual's self-worth (Hayman et al., 2016). Resources include family and friends, but also voluntary and community organisations. The support networks of elderly migrants are however often complex as they can transcend national boundaries (Baldassar, 2014). For retirees in Spain, family often live in the UK and whilst some have retained strong ties these are at a distance resulting in complex caring relationships (Baldassar, 2014). Older British migrants tend to speak little or no Spanish and rarely integrate into the Spanish community (Hall and Hardill, 2016; King et al., 2000). This can leave them unable to access formal Spanish care and support services (Calzada, 2017), resulting in an ongoing reliance on the UK and the negotiation of old age challenges even more complex.

This chapter explores the way in which older British migrants in Spain draw on individual, family and community networks to develop resilience in old age. It presents the findings from a qualitative research study involving narrative interviews with 20 households of older British citizens in Spain. These interviews sought to understand how older British migrants cope with the challenges and complexities of old age, and more specifically how old age

is negotiated in relation to migration experiences and trajectories. In particular, the chapter focuses on the social networks of older migrants, including
friends and family both in Spain and the UK. It explores the ways in which
resilience can be developed through networks in Spain as well as those maintained across national borders through transnational relationships with
family. Following this short introduction, the chapter moves on to explore the
literature on retirement migration, focusing particularly on social support and
its relationship with resilience. It then outlines the methods used in the study,
followed by the findings, drawing on qualitative interviews to explore resilience among ageing British migrants in Spain at an individual, family and
community level.

Growing old in Spain

Old age, especially the onset of the fourth age, brings with it increased vulnerability as people have to cope with multiple challenges often associated
with a decline in health and the resulting need for care or additional support.
Grundy (2006:107) argues that older people can become vulnerable when
their 'reserve capacity falls below the threshold needed to cope successfully
with the challenges that they face'. Resilience, defined as the ability to cope
with and recover from significant adversity or risk, is essential for protecting
against vulnerability (Walsh, 2006). It is therefore not only about the ability
to cope with, but to recover from serious adversity. Resilience can be enabled
at an individual, family and community level (Distelberg et al., 2015) and
includes drawing on reserves of mental and physical health, wealth or other
material resources, formal social protection as well as family, friends and
other community relationships (Walsh, 2006).

Research on resilience in older people has referred to the importance of
family resourcefulness (Walsh, 2012). Family are seen as a crucial source of
emotional, instrumental and financial support in ageing families, especially
during times of crisis (Park and Roberts, 2002; Phillipson et al., 2001). The
most important source of sustained support during old age is a spouse or
partner (Skerrett and Fergus, 2016; Allan, 1989). Lynch (2007) refers to this
type of primary care relations between spouses/partner as 'love labour',
which is emotionally engaged work with high levels of dependency. For
those whose partner has died or is in need of care themselves, primary kin
(children and siblings) are most significant, though for the childless, a niece
or nephew may be important (Allan, 1989). However, most research on
family care relationships assumes geographical proximity and so such intimate and personal care is much more complex following the migration of a
family member (see Baldassar, 2014). Migration can transform the role of
'family' and may result in a weakening of family bonds, or alternatively
may present an opportunity to strengthen and maintain the collective welfare of the family by developing resilience and flexibility across geographical
boundaries (Liu, 2013).

Baldassar (2014) explored transnational care-giving practices, especially how adult children care for (or about) their ageing parents when they live in another country. She posits that 'distance care' is an important part of the migration process and this has been transformed in recent years through the development of ICTs including skype and social media which enable people to stay in close contact at a distance. However, during times of crisis including acute illness, death and dying, a physical co-presence is more acutely required (Baldassar, 2014). Therefore, transnational caregiving is complex and requires families to be mobile and have the resources (money and time) to both undertake and receive visits to deliver hands-on personal care. Distant caregiving also remains a gendered practice, with women feeling a stronger sense of obligation to visit and provide personal 'hands-on care' to elderly parents, although men may be taking a more active role in care giving for elderly parents by communicating electronically (Baldassar, 2008). Baldassar (2014) also identifies that caregiving is rarely a 'one-to one' relationship and instead part of a broader set of social relationships with both formal and informal actors and agencies in the care network.

Whilst family are the most important source of old age care and support, friends are instrumental in providing practical and emotional support for older people and their carers, as well as in supporting people to cope with widowhood (Allan, 1989; Phillipson et al., 2001). Friends are also an important means of socialising and in retaining independence. Friendships can however decline with age, as opportunities for generating and servicing friendships become limited when people become infirm (Allan, 1989). For those who are frail and housebound, neighbours and proximal friends can prove to be especially important during a crisis and can even be an important substitution for family living far away (Gabriel and Bowling, 2004).

The social and support networks of migrants can however undergo massive transformations when they move. Migration therefore involves developing new strategies to maintain existing relationships across national borders, and also establishing new social and support networks. Referring specifically to older British migrants in Spain, research has shown that most migrants move to established British communities in Spain, and choose to live in purpose-built tourist and residential complexes known as 'urbanisations' that are inhabited largely by other British people. They often speak little or no Spanish and do not mix with the Spanish community leaving their social networks comprising only other British people (Hall and Hardill, 2016; O'Reilly, 2000). This can leave them unable to access Spanish care and support services, as administrators and carers often speak little or no English (Hall and Hardill, 2016; Calzada, 2017). On the other hand, moving to an established British community creates numerous opportunities to develop new friendships, take part in new activities and enjoy an active outdoor social life (even if this is contained to other British people) (King et al., 2000). There are a plethora of social clubs catering to the needs of older British people in Spain and these offer a source of enjoyment, as well as access to friends, information and wider support (Betty and Cahill, 1999). As such, retired migrants tend to have strong social networks and are active members of their community (O'Reilly, 2000; King et al., 2000).

However, as noted above, the onset of old age and frailty can lead to a contraction of social networks, especially with friends. The geographical separation of older migrants from their families in the UK also diminishes their access to familial support. This can lead to increased levels of vulnerability whereby the formal and familial resources these migrants can draw on for resilience, especially in the face of a crisis, are minimised. Following the methodology, the remainder of this chapter explores how resilience at an individual, familial and community level is impacted by the intersection of old age and migration.

Methodology

This chapter draws on interviews with older British people in Spain (over 50 years). It only includes those defined as vulnerable i.e. 'whose reserve capacity fall below the threshold needed to cope successfully with the challenges that they face' (Grundy, 2006:107). Most were in 'critical situations' (Hardill et al., 2005) as they had experienced a radical decline in quality of life due to a decline in health or lack of finance and require additional income or support. As such, most of the interviewees were frail, had health problems and many were in need of long-term care. The interviews focused on how they coped with these challenges, including who they turned to for help and support. The analysis draws on the concept of resilience at individual, family and community levels adding further qualitative understanding to the ways in which the challenges associated with older age are negotiated transnationally.

Interviews were conducted with 20 British households, which included a total of 25 older individuals (16 women and nine men). Household rather than individual interviews were chosen as members of a household tend to have a shared life history, so activities and decisions are negotiated jointly (Allan, 1980). By interviewing at a household level these complex household relationships and interactions were explored resulting in richer, more detailed and validated accounts (Valentine, 1999). For 13 households, an individual was interviewed (one was married, 12 were single/widowed), four interviews were with married couples and three were with one older person and other members of their family (usually daughters). Four interviewees had children living nearby in Spain, 11 had children in the UK (although one had little contact with them) and four had no children (although two had close siblings living in the UK). The average age of all interviewees (excluding wider family) was 78.25 years; however, their ages ranged from 51 and 93 years. Only one participant was under the age of 60. The interviewees had lived in Spain for between one to 34 years and therefore captures the problems associated with a recent move to Spain, as well as those who have 'aged in place'.

Households were located in the Costa Blanca (eight households), Costa del Sol (seven households) and Mallorca (five households). Interviewees were included to reflect diverse characteristics in relation to social class, age, marital status, length of time in Spain and household composition. Narrative interviews were undertaken with each household focusing on participants 'lived experience' and personal accounts, focusing on their vulnerability, resilience and social support as told from the perspective of the individuals involved (Dingwall and Murphy, 2003; Lawler, 2002). A broad interview guide was used to lead the conversation between the interviewer and interviewee, but participants were also encouraged to tell their stories and talk about those issues most important to them. The transcribed interviews were coded and analysed using NVivo computer software. A coding framework was used based on the theoretical interests underpinning the study (focused on individual, family and community resilience) (Attride-Stirling, 2001). The research was carried out to the standards set in the ESRC's Research Ethics Framework and the British Sociological Society's Statement on Ethical Practice. In accordance with these guidelines, the research was conducted with the welfare of participants in mind. Only those able to provide informed consent were included. Pseudonyms have been used to protect the identity of individuals.

Exploring resilience in vulnerable older British migrants in Spain

This section draws on interviews to explore resilience among older British people in Spain focusing on how interviewees utilised individual, family and community resources for support during times of crisis.

Individual resilience

Resilience at an *individual* level refers to how an individual's beliefs and values impact on their response to challenging situations (Aburn et al., 2016). Central to this is self-efficacy, i.e. when people believe they have the ability to do well (Distelberg et al., 2015; Bhana and Bachoo, 2011; Seecombe, 2002). In addition, beliefs that are driven by optimism and hope are fundamental to being resilient (Black and Lobo, 2008). Individual resilience can be seen in the following quote by 79-year-old Ida:

> I think life is a series of learning steps, you deal with it as best you can, what happens to you and then you make the most of life. You only get one chance, don't you?

> (Ida, 79, divorced)

Ida had severe health challenges, and as a result intended to return to the UK. Despite living in Spain for 16 years, Ida retained strong ties to the UK (through her own identity as well as her daughter who lived there). She saw the UK as her home indicating a strong sense of 'Britishness'. Despite her health challenges, she was able to remain positive. Ida had lived on her own in Spain for many years and was very independent. However, it was the onset on old age and associated feelings of vulnerability (exacerbated by recently being mugged) that led to her desire to return to the UK.

Studies cite the importance of independence for old age resilience and well-being (Reichstadt et al., 2007; Allan, 1989) and this was also evident in other interviews. For example, Donald who was almost blind and had other health problems, managed to remain resilient by retaining a high level of independence and a positive outlook:

> I don't like to impose myself on people like that. I can look after myself as I have done all my life so I don't have any problems in that respect. I might do when I am about 95 as I am told I will live to.
>
> (Donald, 80, single)

He therefore did not allow his health adversity to impact on his general well-being. His positive outlook meant that he did not view his blindness and other health conditions as significant challenges. At one point in the interview he even commented, '... I am not giving you much am I? [I have] never had any problems!' At an individual level he therefore had developed strong emotional coping strategies. However, he did also acknowledge his wider support network that includes local people whom he draws on as and when he needs them:

> [Friends and other local people] help me across the road. [Friend] made a white stick for me. If I go to [town] the bus drivers are fantastic. They see me and tell me you sit there ... and ask me where I am going, I tell him. They are very good ... The only problem I have is with the shopping, again people are helpful, I can't see the dates, the stuff, I make out my list but then I can't see what I have written so have to get somebody else to read it for me. They will help me in the shops.
>
> (Donald, 80, single)

This indicates how individual resilience cannot be created in a vacuum (Distelberg et al., 2015) and indicates the multidimensional nature of resilience. Individual resilience can therefore be strengthened through wider social support at a community (or family) level (Martin et al., 2015).

Another interviewee, Robin, also displayed high levels of individual resilience. He had encountered severe financial difficulties but had been instrumental in obtaining financial support from his local Town Hall in Spain:

> You get stuff from the Town Hall ... I have just found out that I am entitled now, I can go to town hall, because we are so skint, we can get subsidised housing ... [The information] is there, but you have got to find it, you have got to ask.
>
> (Robin, 62, married)

He is a strong and persistent person, with a strong 'locus of control' (Distelberg et al., 2015), i.e. a belief that he can influence the events that affect him. This was further displayed when he launched an appeal to reinstate his UK disability benefits indicating not only a high level of individual resilience but also an expression of his ongoing rights as a British citizen:

> The daily living allowance. They have come up with this 26 week thing … I have looked online, I have checked … and it all comes back to this European Community which should transfer everything here, but they won't have it. I keep on about it. If people don't ask you never get … I got onto this high guy [an MP] and he's sticking up for me and saying I'm perfectly right.
>
> (Robin, 62, married)

Individual resilience was also displayed in the reasons why people moved to Spain in the first place. Many of the interviewees had moved to Spain when they were healthy and active; others however had moved to Spain with health problems. This included two interviewees with chronic arthritis that was exacerbated by cold weather. For them, moving to Spain was used as a resilience strategy to prolong their quality of life, as Amy explained:

> I had like an arthritis of the muscles and if I had stayed in England with the weather I would have been in a wheelchair by now. So I needed to come down here because the atmospheric conditions are better for some people with this illness.
>
> (Amy, 51, widowed)

Whilst there were no clear patterns in individual resilience in relation to age or the type of challenges encountered, one pattern that emerged from the interviews was high levels of individual resilience among those with little or no family in either Spain or the UK. Whilst some people, like Robin, were married, most of those with high levels of individual resilience lived on their own. In addition, all of the interviewees quoted above were fairly well integrated in Spain; they spoke at least some Spanish and tended to mix with the Spanish community, e.g. had Spanish friends, went to Spanish bars/restaurants and events. This again indicates the multidimensional nature of resilience, as resilience could be enabled by having the confidence and knowledge to be able to interact within diverse networks, which can also facilitate access to formal networks and support as with the case of Robin. This includes ongoing networks and associated rights based in the UK indicating an ongoing sense of Britishness in some of the interviewees. Their experiences also indicates that resilience does not necessarily mean staying independent for as long as possible. Indeed, it demonstrates the importance of local and transnational networks and rights that can be drawn on for resilience in old age (also see Zontini, 2015).

Family resilience

Research has recognised that crises and life challenges have an impact on the whole family and so resources for individual resilience are often identified through the immediate and extended family (Walsh, 2006). Distelberg et al. (2015) identify a range of family resilience indicators that include a family's shared positive outlook (Walsh, 2003; Black and Lobo, 2008), as well as good communication, connectedness and social and economic resources (Walsh, 2003). The skills held by a family, as well as the way in which they communicate, can be linked to their ability to access resources and find solutions (Distelberg et al., 2015). This section explores family resilience focusing on the relationships that the interviewees have with their families, including the transnational dimension of these relationships. It explores how migration has impacted on families, as well as the ways in which family are drawn on for help and support. The family networks of the interviewees are separated into two categories; those with family in Spain, and those whose family live in the UK.

Family in Spain

Four of the interviewed households included people who moved to Spain to join their family who already lived there. One interviewee moved in her forties to help provide care to her grandchildren in Spain, whilst three moved in the fourth age (aged in their seventies/eighties) to receive care from their daughter in Spain. Migration to both *provide* and *receive* care has been noted in prior research (Ackers, 2004); although care is also recognised to include reciprocal relationships involving both the receiving and giving of care (Baldassar, 2014; Zontini, 2015). For example, Wilma and her husband initially moved to Spain to support her daughter (who is married to a Spaniard) and provide care to her grandchildren. Wilma's own mother moved to Spain at the same time and so Wilma also cared for her in the family home until she died. Therefore, when she moved, Wilma was active in the provision of care. As Wilma and her husband became older, the relationship shifted and her daughter and son-in-law began to provide care for them. This was especially the case when Wilma's husband became terminally ill, and after his death, when Wilma's health also declined considerably:

> My daughter was home in the morning and we used to change [Wilma's husband] and wash him and then my son-in-law, the darling, used to come home every evening to help me because I couldn't do it, you see.
>
> (Wilma, 76, widowed)

Wilma's family display high levels of resilience through their connectedness, cohesion and ability to support each other to withstand and rebound from adversity (Walsh, 2012). They engaged in collaborative problem solving as a family and also reciprocal care that included the provision of both elderly and child care. They also rallied together as a family when faced with financial difficulties following the closure of the family restaurant, which included using Wilma's pension to support them financially. On the other hand, they drew on Wilma's daughter and son-in-law's Spanish language and local connections to seek support from the Spanish welfare state when Wilma and her husband needed help. Their resilience as a family was therefore strengthened by its cultural mix, drawing on their strong ties to both the Spanish and British communities. This included receiving some financial help from a British-run voluntary organisation:

> But my parents have got a lot of debts from the restaurant … When you have got a bill for 1,000, 1,500 euros for a funeral we just don't have it. We have got this bill for the dentist now for her teeth, which we just can't pay. We are going to pay it slowly. So [voluntary organisation] have really helped, because they paid for the funeral … .
>
> (Wilma's granddaughter)

Wilma's narrative indicates a high level of role adaptation and flexibility within her family unit, as they drew on the strengths and resources of each family member during major challenges (Martin et al., 2015) as they were able to adapt, re-organise and rebound from bereavement, financial struggles and ill health.

On the other hand, the remaining three older interviewees with family in Spain had moved in their fourth age to *receive* care from their daughter living there. Whilst moving to Spain represents an ability to adapt in response to health challenges, their poor health left them unable to socialise and so they became socially isolated and dependent on their daughter. The severe health problems of these interviewees meant that their family reached a point where they needed additional help with care and so had to search for solutions outside of the family. None of them had the money to pay for private care in Spain, and state-funded care is very limited due to a tradition of family care (Hall and Hardill, 2016). For two of the interviewees, this led to a return to the UK indicating the ongoing reliance that older migrants have on the UK welfare state, especially when they need care and support.

Family in the UK

The remaining 16 households did not have any close family in Spain but most did have family in the UK. They found that when faced with a crisis, the support their family could provide was limited by geographical separation. This included Shirley and Andrew who reached crisis point when Shirley became very ill. Despite having a close relationship with their daughters in the UK, they could only visit for short periods of time and ultimately their only option was to pay for professional care in Spain as Andrew explained:

I had one of my daughters here and she was helping and another daughter came out as well. But they can't be here all the time obviously. She said, dad, you have really got to get some help. Get mum in a home or get someone to come round, she said.

(Andrew, 81, married)

Most of the interviewees retained an emotional relationship with their family in the UK through visits (in both directions when possible), plus phone calls, skype and other forms of electronic communication. However, in the event of a crisis or substantial challenge, family were too far away to be able to offer any instrumental help. As a result, for those who were married, 'couple resilience' (Skerrett and Fergus, 2016) was especially important whereby a spouse or partner was the most important in helping to overcome a crisis or in negotiating the challenges of ageing. Shirley spoke about how dependent she was on her husband. She felt that without him she would not be able to cope so would return to the UK to be close to her children:

I am lucky I have got a husband that … if I was on my own, I would want to sell up and go back to England to live.

(Shirley, 87, married)

Interviewees who were widowed and had no children (or did not have a close relationship with children) found it particularly difficult to access help and support, at both an emotional and practical level. Elsa had severe health difficulties and was house bound. Her husband had recently died and they had no children. Whilst she was close to her sister who lived in the UK, their poor health and the cost of travel meant that Elsa and her sister were unable to visit each other:

My sister, I ring quite often, but she's 82, she's not very well … I'd like to see her but it costs me a lot of money. She is in the UK.

(Elsa, 78, widowed)

Financial constraints can therefore restrict family support and increase vulnerability, especially when relationships are transnational (Zontini, 2015). Financial help was at times offered by children in the UK including helping them with the cost of visiting (or even permanently returning to) the UK. Rachel who is widowed lives alone in Spain, but visits the UK regularly to see her family. She has very little money and so her family pay for her flight:

I went home several times last year. But having said that this is where my family come in … I don't have to worry about the plane tickets. They are very good to me. [Son-in-law] said have you got the ticket he didn't even ask how much it was he just put notes in an envelope and said thanks for coming and that will see to your plane fare. When I go they would be terribly upset if I offered anything for staying there.

(Rachael, 68, widowed)

Therefore, despite geographical separation, some interviewees were able to maintain their family relationships across national borders through visits. It was, however, when a crisis arose or migrants experienced a severe decline in health that family relationships became crucial yet more difficult to maintain. As Ahmed and Hall (2016) suggest, ageing and the resulting dependence, frailty and loss of mobility represents an important structural context that restricts migration experiences. This forces migrants to evaluate their identity as a migrant and choose whether to 'stay or return'. Seven of the interviewees had made the decision to return including Ida who wants to return to the UK to be close to her daughter and sister there. She currently stays in touch by phone and letter but found this was no longer enough:

I am alright on my own for a little while but I need to talk. My [daughter] phones me every day and [other daughter] phones me once a week and I phone her and I do write a lot of letters. I write to my sister regularly, thousands of words I have written over 16 years but she likes to know all the gossip and what I do and everything.

(Ida, 79, divorced)

Therefore, family, especially children (and daughters in particular), were still the most important source of support for many, and as they aged, they wanted to be physically close to them supporting the above notion that old age care and support cannot always be provided at a distance.

On the other hand, those with no family in Spain often spoke about the importance of their friends. Friends were instrumental in providing emotional and practical help, particularly during times of crisis as Audrey explains:

The neighbours kept on coming in to look after [terminally ill husband], because he wasn't eating much by then, making sure he took his tablets … When [my husband] was dying my friends came and stayed with me and I stayed in the A&E and 5 o'clock the next morning he gave up the fight.

(Audrey, 66, widowed)

Amy's friend was able to provide her with practical help by supporting her financially through subsidising her housing, food and bills:

[Friend] is only charging me 300 euros a month and that includes my food, and electric and water. I couldn't stay in the apartment I had, even though the landlord said he would reduce the price.

(Amy, 51, widowed)

Some interviewees considered friends to be as important if not more important than family, especially for those without children. Some even referred to friends as family:

> [Friend in UK] you see, I have known her for over 30 years and she's more like a sister or mother to me.
>
> (Mary, 81, married)

> My second dearest friend, well the one that I am very close to now, is my sister.
>
> (Rachael, 68, widowed)

This indicates the complexity of social networks with a blurring of the boundaries between friends and family (Pahl and Spencer, 2004; Walsh, 2012). Weeks et al. (2001:9) suggest that friends are sometimes becoming 'families of choice' and whilst referring primarily to same-sex relationships they argue that as people have 'the right to define significant relationships and decide who matters and counts as family'. The geographical dispersion of transmigrant families can mean that friends, especially those that are locally based, are drawn upon for support and security (Pahl and Spencer, 2004). This may indicate that friends can be a replacement for family who do not live nearby.

The informal support networks of older migrants can therefore be complex. Informal support, especially through family has been identified as central to resilience in advanced old age (Hayman et al., 2016). Interviewees had to deal not only with the complexities and challenges of old age, but also develop and manage transnational relationships with their family. For most, this involved redefining relationships following migration and developing resilience strategies when faced with a crisis. Although what is particularly evident here is that following the onset of advanced old age, drawing on family during times of crisis was not easy and as Baldassar (2014) has previously noted, geographical proximity is normally required to cope during care crises. Many of the interviewees therefore developed alternative resilience strategies including returning to the UK or moving to Spain to be close to family, or alternatively developing stronger informal networks in Spain including with friends as well as the wider community, which will be discussed next.

Community resilience

Families and individuals do not tend to deal with their crises and challenges in isolation, but are intertwined with and influenced by their community and wider physical environment (Walsh, 2012; Distelberg et al., 2015). Studies indicate that individuals and families with good connections within their community have higher levels of cohesion, have a stronger sense of belonging and so can better adjust to challenges (Martin et al., 2015; Black and Lobo, 2008; McCubbin and McCubbin, 1996). Community resilience indicators include involvement in and support from the community and neighbourhood, as well as access to formal resources including healthcare. Community therefore includes friends, neighbours, voluntary organisations and other natural support systems (Landau, 2010). Therefore, an individual's and family's aptitude for resilience is linked to all of the relationships that embody them and their ecological context (Patterson, 2002; Martin et al., 2015).

Prior research has referred to the strength of the British community in Spain (O'Reilly, 2000; King et al., 2000) and this is enacted through the urbanisations people live in, the places they shop and people they socialise with. Scott (2007) also refers to the significance of voluntary and community organisations for migrant groups in maintaining community as well as transnational ties. The interviewees spoke about the importance of being involved in social clubs and indicated that these facilitate strong networks that provided not only socialisation but also access to information and support:

> We do Irish roll bowls, and we go to a slimming club on a Friday, all be it that it doesn't work very well. Its good any way, it's got a social side to it.
>
> (Victoria, 67, married)

> I am a member of ESRA, the English Speaking Residents Association. I go to their lunches or whatever … It's very helpful for people, I would recommend anybody coming here to join ESRA. They tell us what is going to happen next week.
>
> (Celia, 90, divorced)

British run voluntary organisations also played an active role in the lives of older British migrants in Spain (also see Hardill et al., 2005), and for some were crucial to well-being and accessing old age help and support:

> [Charity volunteer] is fantastic, he gives up his time like everybody else does for [charity] and he takes me [to hospital] every time and without him I would be sunk no doubt about it.
>
> (Donald, 80, single)

Amy, who was the youngest interviewee, spoke fluent Spanish and was well integrated. Despite this, she spoke about the importance of the British community including voluntary organisations as this provided access to help and support during times of crisis:

> We need the [British] charities yes, we always need that because they provide a network of support and they do a lot out here. They have clubs and it's good for people to go to them. If you still want to retain that

link, even if you do not want to live in the UK, you still want to maintain that Englishness. Discuss things that happen in the UK with other people. You want that. Sometimes friends that are English, particularly in times of stress or when you are grieving.

(Amy, 51, widowed)

Amy's resilience largely came from her integration into both the Spanish and British communities and this gave her access to both formal and informal support. However, for interviewees who did not speak Spanish and had no links to the Spanish community, growing old in Spain was more challenging. Navigating care systems was especially difficult leading some to consider returning to the UK:

When I get really old, and I can't do things for myself, then I plan to go back to England. Most of the old people's homes here, apart from the private ones, which I couldn't afford, the majority of the people are Spanish. Whilst I like being on my own, I don't like not being able to communicate easily.

(Robert, 72, divorced)

This demonstrates the importance of belonging to both the British and Spanish communities in Spain, especially for migrants who have no family nearby. Most older British migrants in Spain are part of the British community in Spain and this is central to their ongoing identity and relationship with the UK that encompasses social networks as well as rights and entitlements (O'Reilly, 2000; Hall and Hardill, 2016). This ongoing Britishness and lack of ties in Spain can limit their ability to access formal support, including health, care and financial support services that are available to all Spanish residents (including British nationals) but often requires a good understanding of the Spanish language.

Conclusion

This chapter has focused on a sub-set of British people living in Spain; those who have become older and encountered a severe challenge or difficulty. One of the most significant and common challenges in old age is a decline in health and the resulting need for care. Other (and often associated) causes of crisis include caring for a spouse/partner, bereavement and financial difficulties (including the need to pay for care). Responses to these crises differed among the households interviewed. To understand vulnerability and coping strategies, this chapter drew on a lens of resilience. Recognising that resilience is not just developed at an individual level, it also explored family and community resilience strategies and indicates that resilience is multifaceted involving the interplay of all three levels (Distelberg et al., 2015). It has helped to develop a qualitative understanding of how older migrants navigate the challenges of old age by drawing on both locally based and transnational networks of support including family, friends and community and voluntary organisations.

Resilience is also about forward planning and the development of protective resources to shield against risks (Hayman et al., 2016; Walsh, 2012). The research presented here indicates that older British migrants who have planned ahead and mobilised preventative resources are the best placed to deal with adversity. This includes being adaptive and flexible in response to age-related challenges, which can include making the (often difficult) decision to return to the UK to access informal and formal support. Making the decision to return can involve reassessing ideas around belonging and home, including their identity as a migrant (Ahmed and Hall, 2016) as well as their ongoing 'Britishness'. Interestingly, three of the interviewees moved to Spain in their fourth age to receive care from their daughter. However, in each of these cases, the older interviewees had no other support networks around them when they moved to Spain and so became dependent on their daughter for all of their support needs. This lack of wider support could be attributed to the fact that they moved in the fourth age and with existing care needs. Therefore, this prevented them from developing social networks and integrating into the wider British and/or Spanish community. As with Martin et al. (2015), this therefore indicates that an individual's and family's aptitude for resilience is mediated by other relationships with the community which they inhabit including the wider family, friends, and other services/systems.

Other interviewees who moved when they were active and healthy were able to develop strong networks of support in Spain, including close friendships that provided emotional and practical support, as well as socialisation. However, the research did find that older, frailer interviewees had fewer friends and joined in with fewer social activities. Many had lost close friends over the years; either because they had died or returned to the UK. Friendships therefore tend to decline in old age and for most, family remain the most important source of support (Allan, 1989). For some of the interviewees, this triggered a return to the UK to be close to family. However, those without close family in the UK often turned to the wider community including voluntary organisations in Spain. British-run voluntary organisations that operate in Spain (e.g. Age Concern Espana, Age Care Association) are often crucial in providing support during old age. These voluntary organisations can span the boundary between formal and informal support by not only providing information and advice, but also emotional and practical support. Hall and Hardill (2016) have argued that they can play a similar role to friends especially when providing emotional support during times of ill health or bereavement. They can also support access to health, care and other support services in Spain, by providing information in English, translation at and transport to medical appointments.

The chapter demonstrates therefore not only the importance of strong and varied support networks in old age, but also that the complexities of growing old as a migrant means that these networks often need to be maintained both locally and transnationally. For older migrants, being part of the British community in Spain and also maintaining transnational ties with family in the UK are crucial to positively responding to and overcoming adversity. However, resilience is arguably stronger for those who are integrated into Spanish community as speaking Spanish and understanding cultural norms can be crucial to accessing support from Spanish social services and therefore building resilience. This may be especially important in light of policy changes that may arise since the Brexit vote. The rights of British pensioners to both reside and access health and social care in Spain may not continue once the UK leaves the EU. Being resilient in response to Brexit will require additional resources, social networks and an internal strength that will need to be explored in further research.

References

Aburn, G., Gott, M. and Hoare, K. 2016. 'What is resilience? An Integrative Review of the empirical literature'. *Journal of Advanced Nursing*, 72(5): 980–1000.

Ackers, L. 2004. 'Citizenship, migration and the valuation of care in the European Union'. *Journal of Ethnic and Migration Studies*, 30(2): 373–396.

Ahmed, A. and Hall, K. 2016. 'Negotiating the challenges of ageing as a British migrant in Spain'. *GeroPsych*, 29: 105–114.

Allan, G. 1980. 'A note on interviewing spouses together'. *Journal of Marriage and the Family*, 42: 205–210.

Allan, G. 1989. *Friendship: Developing a Sociological Perspective*. Hemel Hempstead: Harvester Wheatsheaf.

Attride-Stirling, J. 2001. 'Thematic networks: an analytic tool for qualitative research'. *Qualitative Research*, 1(3): 385–405.

Baldassar, L. 2008. 'Missing kin and longing to be together: emotions and the construction of co-presence in transnational relationships'. *Journal of Intercultural Studies*, 29(3): 247–266.

Baldassar, L. 2014. 'Too sick to move: distant "crisis" care in transnational families'. *International Review of Sociology*, 24(3): 391–405.

Betty, C. and Cahill, M. 1999. 'British expatriates' experience of health and social care services on the Costa del Sol'. In *Into the Margins: Migration and Exclusion in Southern Europe* edited by F. Anthias and G. Lazaridis, 83–104. Ashgate: Aldershot.

Bhana, A. and Bachoo, S. 2011. 'The determinants of family resilience among families in low- and middle-income contexts: a systematic literature review'. *South African Journal of Psychology*, 41(2): 131–139.

Black, K. and Lobo, M. 2008. 'A conceptual review of family resilience factors'. *Journal of Family Nursing*, 14(1): 33–55.

Calzada, I. 2017. 'Social protection without borders? The use of social services by retirement migrants living in Spain'. *Journal of Social Policy*, doi:10.1017/S0047279417000101.

Dingwall, R. and Murphy, E. 2003. *Qualitative Methods and Health Policy Research*. New York: Aldine de Gruyter.

Distelberg, B., Martin, A., Borieux, M. and Oloo, W. 2015. 'Multidimensional family resilience assessment: the Individual, Family, and Community Resilience (IFCR) Profile', *Journal of Human Behavior in the Social Environment*, 25(6): 552–570.

Gabriel, Z. and Bowling, A. 2004. 'Quality of life from the perspectives of older people'. *Ageing and Society*, 24(5): 675–691.

Giner, J., Hall, K. and Betty, C. 2015. 'Back to Brit: retired British migrants returning from Spain'. *Journal of Ethnic and Migration Studies*, 42(5): 797–815.

Grundy, E. 2006. 'Ageing and vulnerable elderly people: European perspectives'. *Ageing and Society*, 26: 105–134.

Hall, K. and Hardill, I. 2016. 'Retirement migration, the "other" story: caring for frail elderly British citizens in Spain'. *Ageing and Society*, 36(3): 562–585.

Hardill, I., Spradbery, J., Arnold-Boakes, J. and Marrugat, M.L. 2005. 'Severe health and social care issues among British migrants who retire to Spain'. *Ageing and Society*, 25(5): 769–783.

Hayman, K., Kerse, N. and Consedine, N. 2016. 'Resilience in context: the special case of advanced age'. *Aging and Mental Health*. doi:10.1080/13607863.2016.1196336.

King, R., Warnes, T. and Williams, A. 2000. *Sunset Lives: British Retirement Migration to the Mediterranean*. Oxford: Berg.

Landau, J. 2010. 'Communities that care for families: the LINC model for enhancing individual, family, and community resilience'. *American Journal of Orthopsychiatry*, 80(4): 516–524.

Lawler, S. 2002. 'Narrative in social research'. In *Qualitative Research in Action*, edited by T. May, 242–258. London: Sage.

Liu, J. 2013. 'Ageing, migration and familial support in rural China'. *Geoforum*, 51: 305–312.

Lynch, K. 2007. 'Love labour as a distinct and non-commodifiable form of care labour'. *Sociological Review*, 55(3): 550–570.

Martin, A., Distelberg, B. and Elahad, J. 2015. 'The relationship between family resilience and aging successfully'. *The American Journal of Family Therapy*, 43(2): 163–179.

McCubbin, M. and McCubbin, H. 1996. 'Resiliency in families: a conceptual model of family adjustment and adaptation in response to stress and crises'. In *Family Assessment: Resiliency, Coping and Adaptation: Inventories for Research and Practice*, edited by H. McCubbin, A. Thompson and M. McCubbin, 1–64. Madison: University of Wisconsin System.

O'Reilly, K. 2000. *The British on the Costa del Sol: Transnational Identities and Local Communities*. London: Routledge.

Pahl, R. and Spencer, L. 2004. 'Personal communities: not simply families of "fate" or "choice"'. *Current Sociology*, 52(2): 199–221.

Park, A. and Roberts, C. 2002. *The Ties that Bind, British Social Attitudes, the 19th Report*. Aldershot: Ashgate.

Patterson, J.M. 2002. 'Understanding family resilience'. *Journal of Clinical Psychology*, 58(3): 233–246.

Phillipson, C., Bernard, M., Phillips, J. and Oggy, J. 2001. *The Family and Community Life of Older People: Social Networks and Social Support in Three Urban Areas*. London: Routledge.

Reichstadt, J., Depp, C.A., Palinkas, L.A., Folsom, D.P. and Jeste, D.V. 2007. 'Building blocks of successful aging: a focus group study of older adults' perceived contributors to successful aging'. *American Journal of Geriatric Psychiatry*, 15: 194–201.

Scott, S. 2007. 'The community morphology of skilled migration: the changing role of voluntary and community organisations (VCOs) in the grounding of British migrant identities in Paris France'. *Geoforum*, 38: 655–676.

Seecombe, K. 2002. '"Beating the odds" versus "changing the odds": poverty, resilience, and family policy'. *Journal of Marriage and Family*, 64(2): 384–394.

Skerrett, K. and Fergus, K. 2016. *Couple Resilience: Emerging Perspectives.* New York: Springer Publishing Company.

Thurley, D. 2014. *Frozen Overseas Pensions.* [Online] Available at: http://frozenbritishp ensions.org/wp-content/uploads/2015/07/SN01457-8.pdf [Accessed 13 January 2017].

Valentine, G. 1999. 'Doing household research: interviewing couples together and apart'. *Area*, 31: 67–74.

Walsh, F. 2003. 'Family resilience: a framework for clinical practice'. *Family Process*, 42(1): 1–18.

Walsh, F. 2006. *Strengthening Family Resilience*, 2nd edn. New York, NY: Guilford.

Walsh, F. 2012. 'Successful aging and family resilience'. In *Emerging Perspectives on Resilience in Adulthood and Later Life*, edited by B. Haslip and G. Smith, 153–172. New York, NY: Springer.

Weeks, J., Heaphy, B. and Donovan, C. 2001. *Same Sex Intimacies: Families of Choice and Other Life Experiments.* London: Routledge.

Zontini, E. 2015. 'Growing old in a transnational social field: belonging, mobility and identity among Italian migrants'. *Ethnic and Racial Studies*, 38(2): 326–341.

11 Returning at retirement

British migrants coming 'home' in later life

Katie Walsh

There is now a burgeoning interdisciplinary literature on migration in later life, reflecting the increasing numbers of people living longer and negotiating ageing within transnational social fields (e.g. Horn and Schweppe 2015; King et al. 2017; Näre et al. 2017). These studies have mainly focused on two distinct groups of older migrants: firstly, labour migrants 'ageing-in-place', for whom studies highlight the economic deprivation and social marginalisation experienced by many in Europe (Buffel 2015); and, secondly, more affluent retirement communities of lifestyle migrants (e.g. Benson and O'Reilly 2009). British migrants have been considered within the second of these two literatures, as retirees, with studies highlighting migration to southern Europe (King et al. 2000; O'Reilly 2000; Oliver 2008), as well as Thailand (Botterill 2016). Scholars have noted both the 'active ageing' strategies that international retirement mobilities afford Britons, while also revealing the vulnerabilities of advanced older age (illness, bereavement, etc.) that sometimes lead to return (e.g. Giner-Montfort, Hall and Betty 2016; Hall, Betty and Giner 2017; Hall and Hardill 2014; Oliver 2008). These studies (conducted prior to the UK's EU membership referendum) suggest that British retirees living in such communities mostly do not intend to return to the UK at all (Giner-Montfort, Hall and Betty 2016; Warnes et al. 1999), even in terms of the repatriation of their body after death (O'Reilly 2000). However, ageing can often decrease resources for independent living and, thereby, encourage return to access familial support or health and social care services in later stages of older age (Hall, Betty and Giner 2017; Hall and Hardill 2014; Giner-Montfort, Hall and Betty 2016; Oliver 2008; Percival 2013). In the context of their research on British retirement migrants returning from Spain, for example, Giner-Montfort, Hall and Betty (2016, 798) argue that research has focused too much on the positive lifestyle outcomes for retirees moving to Spain and the positive impact of their settlement for the host economy, constructing an image of British retirement migrants that obscures any consideration of the 'vulnerability, isolation or bereavement that may lead to a return move'.

The numbers of Britons returning are difficult to capture, like the data on those still living outside Britain (Giner-Montfort, Hall and Betty 2016; Finch, Andrew and Latorre 2010). Perhaps this contributes to the invisibility of

British returnees since there is relatively little discussion of British return migration, beyond this literature on those returning in advanced older age from retirement communities identified above and a small number of studies on the return of emigrants from colonial service and/or settlement (Harper 2005; Knowles 2008). However, return migrants also make 'surprising' appearances in other studies. Hammerton and Thomson's (2005) research on 'ten pound poms' *emigrating* to Australia with the anticipation of permanent residence, includes discussion of the unexpectedly high number who returned (some of whom then re-emigrated), as well as noting how many more 'longed' to return but found themselves unable to due to the expense. Rogaly and Taylor (2009) also found that among their white working-class interviewees on a social housing estate in Norwich, England, many of those in their sixties and seventies had been based abroad in the British military as service men and women in the last years of colonial rule in, among other places, Yemen and Malta. Together these studies provide rich evidence of return migration being a taken-for-granted dimension of British culture and society, both in the UK and beyond. In British migrant communities, return is a topic of frequent discussion and sometimes uncertainty, with the option of mobility being a significant resource in ongoing lifestyle, family and career projects (e.g. Conway and Leonard 2014; O'Reilly 2000; Walsh 2018).

Existing research on British return migration is by no means exhaustive then in terms of understanding the patterns and experiences of return, since British migration itself is so varied. In this chapter, I therefore aim to contribute to these debates by focusing on the narratives of three interviewees who were living in Dubai, United Arab Emirates, for several decades as British professional and/or 'middling transnational' (Conradson and Latham 2005) migrants and are now, in their seventies, returned to the UK. Professional migrants might also be understood as highly skilled migrants, since they are either university graduates or with equivalent experience, skills, and status in their chosen sector. 'Middling transnationals' are also often highly skilled middle-class migrants, but their skills may not be recognised or rewarded so strongly by the global economy. They may also work in either lower-paid graduate sectors, especially education (including English-language tuition) and journalism, or, alternatively, have pursued non-graduate careers in the retail, hospitality, health, and tourism sectors. The spatial trajectories of both professional and middling transnational migrants' careers may involve multiple 'returns' during their working lives for periods between contracts in different countries or they may circulate direct from one country to another without returning in between. Indeed, they may not consider their global mobilities as *migration* at all, describing it as a *temporary* sojourn, secondment, expatriation, or 'posting' in line with the way in which skilled migrants have been labelled 'skilled *transients*' (Findlay 1988). The inter-company-transferees in Beaverstock's (2005) study, for example, had often had several international secondments in their career and were 'posted' from London to New York's financial district, with a clear sense that this would fast-track their career back in London. However, there is also evidence that some highly skilled British migrants do not anticipate returning at all (Harvey 2009) and, moreover, may later delay or resist returning more permanently at retirement, instead relocating to become part of these lifestyle migrant communities described above (O'Reilly 2000).

Yet, many do return to the UK for their retirement and it is this return in later life that is the focus of this chapter. The accounts I analyse here have been strategically selected from among a wider ongoing project about British return migration, a project that encompasses return from other countries (see Walsh 2016). This is a project that I began in 2010 and am, sadly, yet to complete for reasons including the difficulties of recruiting returnees who, unlike British migrants, are not either residentially co-located or heavily involved in clearly demarcated institutional lives. Nevertheless, strategic snowball sampling through my colleagues, friends and family has allowed me to include 20 individuals and couples so far, encompassing returnees with diverse biographical experiences of residence and mobility, and aged from late sixties to early nineties. Later life is explored in this project not as a fixed biological stage, but as a process that intersects with other events, trajectories, and social processes, such as, of relevance here, international migration and settlement. Indeed, a critical perspective on 'older age' focuses on its meaning as a socially constructed and embodied subjectivity, rather than a *chronological* marker (Hopkins and Pain 2007). While I focus on return in this chapter as being a transition associated with retirement as a particular life stage, I draw analytically on a life course perspective that recognises that: 'rather than following fixed and predictable life stages, we live dynamic and varied life courses which have, themselves, different situated meanings' (Hopkins and Pain 2007). The primary method employed in this study is in-depth interviewing, akin to 'subject-oriented' life histories, in which narration is understood as a practice of self-construction in the present, rather than a simple description of the past (Hollway and Jefferson 2000). Selecting the stories of three individuals to explore here in more depth, allows me to nuance depictions of their privileged working lifestyles enjoyed in the Gulf with the *vulnerabilities* that arise in this later phase of life associated with ageing, retirement and return.

King et al. (2017) argue that we need to counter the dominant trope of vulnerability that surrounds existing studies of older migrants to recognise their active, positive and creative ageing strategies. For British migrants in later life, it is necessary to acknowledge both the privilege of whiteness that has shaped their opportunities and everyday lives as migrants, while also being mindful of the limitations on their agency in later life and any consequent vulnerabilities. As we have outlined in the introductory chapter to this volume, collectively British migrants can be understood as *relatively* privileged, their global mobilities co-constructed through the operation and practices of middle-class whiteness (Benson and O'Reilly 2009; Fechter and Walsh 2010; Knowles and Harper 2009; Leonard 2010; Walsh 2018). As Caroline

Knowles (2005, 107) argues: Empire survives as a feeling of choice and opportunity, (divergent) forms of entitlement, facilitated by a (racialised) geography of routes already carved out and traversed by others. This privilege does not simply disappear in later life or upon return. Indeed, Knowles (2008, 169, 170) demonstrates how whiteness is 'comported on the scenes of everyday life' by exploring how British return migrants to rural Devon, 'bring back artefacts, memories and lives cast in colonial circumstances'. Yet, classed and racialised privilege is dynamic and intersectional in the enactment of British migrant subjectivities.

To avoid homogenising British migrants, this chapter explores those 'vulnerabilities' arising from the intersection and coproduction of whiteness with older age. The article first introduces the broader research project on British return migration within which these three interview narratives with Gulf returnees arose. I then proceed to introduce the informants and contextualise their migration in the earlier phase of urbanisation and development of the UAE. The next section argues that due to their migrant status and residence in the UAE without citizenship, as they approach retirement age their right to continued settlement is called into question, irrespective of their length of stay. To live and work in Dubai, every migrant, even the highly skilled, requires sponsorship from an Emirati national. Their sponsor (*Kafeel*) is usually their employer or that of their spouse or parent. As a result of this system of managing migration, the meaning of 'settlement' in this region is always ambiguous and contested in terms of permanence for those without citizenship (Mohammad and Sidaway 2016). I show how the possibilities of negotiating this legislation in later life are informed primarily by residents' chronological age, but also by their nationality, occupational sector, wealth, and visa status (e.g. as employees or investors). I argue that Britons experience a varied sense of agency in relation to their return after retirement, with some able to exercise more choice and others feeling forced to leave the UAE reluctantly or involuntarily. The subsequent analysis section then explores their vulnerabilities back in the UK as British returnees negotiate the impact of ageing on their social lives and health. I conclude that analyses of British migration need to be more sensitive to the impact of ageing, a process that brings into question their privilege in ways that scholars, and they themselves, may not anticipate.

British migration to the UAE in the 'early days'

British migration to the UAE has a relatively recent history. A series of treaties from 1820, culminating in the Exclusive Agreement of 1892, established the area of the Arabian Peninsula where the UAE is located in a formal 'protectorate' relationship with imperial Britain (Heard-Bey 1982). However, this imperial relation was designed primarily to protect British interests in the Indian sub-continent by safeguarding the sea routes and the establishment of supremacy over other European powers in the region, while internal affairs in the 'Trucial States' remained in the hands of individual rulers (Davidson 2005). As such, this region did not become a destination for British settler migrants over this early period. Indeed, there was not even a British, as distinct from an Indian, representative in the Sharjah-Dubai area until 1954 (Coles and Walsh 2010). Following the discovery of oil in 1966, economic development started to increase migrant labour to the UAE at all skill levels, but the British community remained very small: the first census in 1968 recorded fewer than 400 Britons in Dubai (Coles and Walsh 2010). They consisted mostly of managerial staff in banks, trading, shipping and oil companies, town planning, and public utilities, together with a smaller number of elite British officials and advisors to the Dubai government, and teachers and health professionals to support the families accompanying them (Coles and Walsh 2010).

Following independence in 1971, the urban expansion of Dubai, and the resulting need for skilled migrants, continued steadily through the 1970s and 1980s. This was reflected in the increasing number of middle-class[1] British migrants who moved there to take up employment, especially engineers and those with technical expertise who arrived to build major infrastructure projects: the deep-water port, the new airport, roads, water and sewage systems, and Jebel Ali harbour (Coles and Walsh 2010). It was at this time that Allan Findlay (1988) identified the changing structure of British international migration, with a quantitative study that demonstrated settler emigration to Canada, Australia and New Zealand was declining in significance in relation to the temporary emigration of skilled migrants: 'professional and managerial staff, taken as a proportion of all actively employed emigrants, increased steadily from 37% of the total in 1973 to 59% in 1985' (Findlay 1988, 402). Many of these Britons, were headed to the oil-rich Gulf states, contributing to his observation that from 1983 to 1985 the Middle East displaces the former commonwealth countries as the most important destination for British migrants for the first time in many centuries (Findlay 1988, 406). The privilege of this community was not only linked to their skilled occupations per se, but also from the close relationships with the Emirati nationals that emerged as a result:

> British expatriates viewed Dubai as a safe, relatively easy and tolerant place to live, in contrast to nearby countries such as Aden and Saudi Arabia. They identified strongly with the place: many were implementing exciting and professionally challenging projects that contributed directly to Sheikh Rashid's vision for the future of the state and, as ruler (1958–90), he attracted unusual loyalty and respect from the British expatriate population. Western expatriates in Dubai then were needed and felt welcome.
>
> (Coles and Walsh 2010, 1321)

All three of the British returnees featured in this article arrived at this time of expansion of the British community, as part of the recruitment of skilled migrants in response to broader processes of urbanisation and development in Dubai.

Being male, a civil engineer by profession, and circulating internationally on short contracts, James is the individual migrant among this group who most fits Findlay's (1988) observation of the movement of 'skilled transients' to the Middle East in the 1970s. In his interview, James repeatedly highlighted his consultancy career and the international mobility that resulted, mapping his narrative chronologically in terms of contracts and their locations. He worked also in Nigeria, Oman, India, Iceland, Ireland, Greece, Denmark, and Sri Lanka during his lifetime, but it was the UAE where he settled for several long periods. His narrative drew attention to his responsibility for large-scale infrastructural projects:

> In the early days, it was the road to Hatta, the road to Sharjah, big stuff. Have you heard of Big Red? Well one of my earlier roads was the road that goes past Big Red to Hatta, goes from the Dubai Country Club, well what was, this is the 70s, late 70s, that road, went through the mountains, Madam roundabout, across the gravel plain, through the mountains to Hatta and then it extends through to the far coast.
>
> (James, 2017)

The challenge of the engineering projects he was offered in the UAE brought James to return in 1992, for seven years, and then once more, in 2003, for 15 successive contracts. He was employed on expatriate contracts from Gulf-based international companies, as a Chief Resident engineer on several highways, as well as bridges, flyovers, marinas, and golf courses.

Linda was also part of this early wave of British migrants to post-independence Dubai, arriving with her husband in 1981, in her early forties, after first living in Cyprus and Bahrain. As an accompanying spouse, she didn't need to work and had no visa to do so, but she later divorced and, in an effort to stay on in the Gulf, moved into administrative managerial positions in the business sector. The privilege of the lifestyle that Britons experienced in Dubai at this time is evident from Linda's account of their leisure activities, domestic help, and her own employment:

> I don't regret, ever, going to live in the Middle East, because it was such a fun place to be, especially in the early days when it was a small place really, we made our own entertainment. I used to be in the motor club, we used to rally together. I was the navigator and we used to go out into the desert and we used to go over the sand dunes, it was very exciting. I had nice homes, which the company provided, and a nice salary, no tax, and home help, so I didn't have to worry about doing the housework while I was at work. I'm still in touch with most of my staff. I had about twenty people reporting to me and, even though we were working long hours, it was a real bond – it was exciting – all different nationalities, that was a good thing. That's a very precious memory, very precious. Very sad that I'm not there: I'd like to turn the clock back thirty years.
>
> (Linda, 2017)

Patricia also arrived in the late 1970s as a British teacher from London. Her contract was extended repeatedly and she stayed in Dubai for 35 years. With her husband, Patricia invested in an apartment in one of the new mega-developments in Dubai and they now use this as a second home to spend periods of the English winter in a sunnier climate.

This section has introduced the three transnational British migrants whose retirement and return is discussed in this chapter. While a transnational orientation is prevalent among British migrants in Dubai, with most investing in, and maintaining, property in the UK, at the same, British migrants form attachments to the city, especially those who, like James, Patricia and Linda, lived there over decades (Walsh 2018). In spite of their privileged residence in Dubai, their 'settlement' as working migrants is determined by their visa status as employees and investors and they have not gained citizenship. The Emirates has been very strict in granting citizenship since the 1990s and, while the requirements are unspecified, Ali (2011) reports that they are thought to be: over 30-years' residence, being Muslim, of Arab descent and an Arabic speaker, as well as having a clean police record, good academic qualifications, a certain level of bank savings, and a degree of personal influence (*wasta*) to lobby a sponsor. Among elite south Asians in the UAE, there is a community of wealthy Indian merchants who have achieved citizenship status and the right to the permanent settlement that it confers (Vora 2013). In contrast, British residents do not speak of UAE citizenship at all, except to note their vulnerability to certain legislative and bureaucratic processes as non-citizens (Walsh 2018; this also contrasts with Bahrain). As such, retirement in later life brings with it the potential to lose their right of residency. The next section explores the impact of the transition to retirement.

Working visas, retirement, and return from the UAE in later life

The UAE Federal Law for Pension and Social Security stipulates the retirement age is 60 or 65 depending on the sector. For those working in Dubai International Financial Centre, legislation relating to retirement age does not apply (Reuters 2016). However, for most employees approaching 65 years of age they must seek approval from the Ministry of Labour to continue working. As such, for the three British migrants whose stories are featured in this chapter, reaching retirement age in Dubai required them to negotiate their right to continue living there. As Linda (70) put it, 'there is some lee-way' (interview, 2017), with one-year labour contracts renewable each year following for those in particular sectors with a skill-set and local experience that makes them difficult to replace. Alternative routes to staying in the UAE past retirement age include: (a) obtaining sponsorship through a family member; (b) accessing an investor visa; or (c) becoming a resident in a continuing care retirement community (Reuters 2016). There is also some ambiguity for those who have invested in property. Yet, all these alternative routes to retirement in the UAE are expensive and, therefore, likely only to be available to those who either have significant inherited wealth or have been working in highly skilled occupations in the Gulf for many years allowing them to build up the financial reserves to explore these options. A further option would be to become a dependent of a working-age child in Dubai, but extended families of this sort are rare among British nationals in Dubai since most children return to the UK for higher education and may then seek employment there or elsewhere.

The legislative environment of the UAE is dynamic as they adjust to the increase in their population, largely from labour migrants. British migrants narrate a loss of status of the British community in relation to the Emirati nationals since independence, with an impact on retirement specifically: prior to the influx of Britons in 2000s, several Britons who had been highly influential and contributed to the infrastructural development of the UAE, for instance the port at Jebel Ali, had been granted special permission to stay on past retirement. As Patricia explains:

> They've all died off now, but they did stay on. We spoke to a chap who built the airport, he was The Airport Man, and he said 'they're very kind, they let me have an office there, but I don't really do anything.' And I think this was the same with all of them. And some of them who actually stopped working, I think it was the customs man we heard about, they actually gave him a house and said 'here you are,' and he stayed on until he died.
>
> (Patricia, 2017)

Having observed this change, Patricia was aware that her own experience was 'lucky'. In her fifties, Patricia decided to start a business and the success of this investment, as well as her investment in property, enabled her to stay on into her seventies, accompanied by her husband. This would not have been possible had she remained with the visa of an employee, but having started her own business and, with the support of her *Kafeel* (sponsor), turned it into a Limited Liabilities Company, Patricia reflected:

> We have the status of investors, so we could stay as long as we like: there is no age limit. I was very lucky: I've got an extremely good sponsor. So, our residency permit lasts three years instead of two, and we have no labour card so we don't have to go through all the nonsense of renewing a labour card. So, that's made life much easier. We were fortunate that, because we were self-employed, we had the choice of what we did in Dubai.
>
> (Patricia, 2017)

In contrast, James did not have the status of an investor and, in spite of his contribution to the development of Dubai as a civil engineer, nor did he have the *Wasta* or status of 'The Airport Man' described above. Nevertheless, James was able to continue working in the Emirates until he was in his mid-sixties since, as an engineer, his skills were in high demand. Yet, even then, the global financial crash reduced the scale of the construction projects he was employed in before he was ready to retire. As a skilled British migrant, James was not vulnerable to the financial crisis to the extent that low-income migrant workers were in the UAE. Research by Michelle Buckley (2012) details how low-income Indian migrant workers were viewed as disposable labour, with companies withholding wages and organising mass deportation flights to remove the workers no longer needed when contracts stalled with financing delays. Yet his age did increase James' vulnerability in comparison with some younger British migrants who were able to more easily change or accrue new contracts:

> End of 2009, they said 'sorry, we don't have any work.' They were actu-ally laying-off directors and I was, I don't know, how old was I then? I was mid-sixties then. I was over the retirement age when they employed me. If they consider that they need you, and they don't have anyone else local, then the employer can pay the government, I don't know how many thousand to extend your visa. Until they had no more work, well I was on a tiny little job and they couldn't justify me being there. I found I was doing parking bays and a little bridge with a Chinese contractor near Big Red and, basically, I was too expensive I think. They must have laid-off sixty, seventy per cent of staff, but I received gratuities. Anyway, I was old enough.
>
> (James, 2017)

James' experience reminds us that, in spite of their privilege, British migrants' everyday lives are also haunted by financial precarity without the rights to residence or welfare (see Botterill 2016 on the impact of the global economic downturn on British retirees in Thailand). Not wishing to retire at this age, since 'the work was interesting', he chose instead to relocate for three years to supervise a section of 'the first ever motorway type road in Sri Lanka'.

In contrast to Patricia and James, Linda felt she was 'forced' to return. When Linda reached sixty years of age, she was invited to continue working for a limited period as a consultant for specific projects:

> I took that, because I really didn't want to leave. I had relatives living there, my nephew was married and they had young babies, so it suited me to stay on. I stayed on for another eighteen months or so, until I was about 62, or something like that, 62, 63. I did several really interesting projects, which I really enjoyed, then the time came, and they said: 'well, we can't really keep you on anymore.' So, that's the time when I had to retire formally.

Hall, Betty and Giner (2017) note that the level of choice and control that British retirement migrants feel in relation to their decision to return varies, with some facing unwanted, involuntary returns that leave them feeling vulnerable and unhappy. Their observation resonates with Linda's account of her own return:

> I didn't want to come back to England. I'd bought a property in Dubai, hoping to stay there. The only way that you could stay there after your retirement, because that's when your residence permit expires, was to be a property owner, so I thought, 'I would like to stay in Dubai, so the only way I can do that is to buy a property.' It was a real disappointment when it, sort of, went wrong [the project was not completed]. I actually contacted the ruler's office and tried to do something but, in the early days when expatriates could buy property, there were no rules really and we didn't know what we were entitled to, or how protected or unprotected we were. By this time I was on my own, I didn't have a partner, so I felt quite vulnerable, as you can imagine.
>
> (Linda, 2017)

Linda's nationality allowed her to stay in Dubai for a further two years after she stopped working, since she was able to complete 'visa runs' to renew the 30-day tourist visa available to those with UK passports. However, without a sponsor, renting property was not a choice available to Linda so she lived with family and friends, eventually receiving a refund of the money she had invested. This was fortunate as after job-losses, divorce and other crises, some Britons in Dubai find themselves in such acute financial need they are forced to approach the British Community Assistance Fund to help them return to the UK. Nevertheless, the experience of buying property changed her desire to stay in Dubai:

> I decided that I did not want to invest any more money in that country, even though I might have been offered a watertight project or something. I decided then that I would come to England and the only safe place to buy a home, or to put your money into a home, was England, so that was my feeling at the time.
>
> (Linda, 2017)

The association between home ownership and the production of middle class identities has long been evident in English culture (Saunders 1989), but is being enhanced by global neoliberal dynamics in which the financialisation of home plays a key role (Forrest and Hirayama 2015). Here though, there is a strong sense in Linda's narrative that the link between property investment and home is more about *ontological* security (Saunders 1989) and the desire for a continuity of settlement in later life.

This section has explored narrative evidence to consider how their status as migrants is experienced by British residents in Dubai, specifically in relation to their sense of agency in terms of retirement and return as they enter their sixties. It demonstrates that individuals residing in the Gulf have varied experiences of retirement, even among the British middle-class migrants who have made Dubai their home for decades. While chronological age is given more determinacy that in the UK due to the Kafala sponsorship system through which migration is managed in the UAE, variation still arises from the occupational and financial resources migrants have to navigate this legislation. While all are potentially vulnerable from their migrant status, some are more able to choose the terms of their return in later life. Most notably, those with investor status due to their business and property ownership, are able to significantly delay return and continue to have a second home in the UAE. While for others, the social meaning of ageing encoded in migration management legislation limits their employment opportunities and, thereby, their rights to residence. The next section further explores the vulnerabilities of return by focusing on experiences back in the UK.

Experiences of British returnees 'settling' back 'home'

Existing research on British return migration suggests that 'coming home is harder than you imagine', evoking a range of, sometimes contradictory, emotions (Hammerton and Thomson 2005, 304). Linda's narrative certainly revealed such a struggle. Returning reluctantly, she attempted to make her new home in Yorkshire where she had grown up, and where her siblings still lived. This had also been a UK base during residence in Dubai, as she had owned a property there (before selling it to invest in Dubai) and visited annually. Nonetheless, now in her early seventies, Linda explained:

> It took me an awful long time to adjust to life in England, because I'd been living abroad for more than thirty years, I felt like a complete, um, foreigner really, coming back to England: I didn't know what people were talking about, I didn't know the rules … So, I basically came back to my roots. I feel safe now, but I'm not one hundred per cent settled. It's not out of the question for me to leave and go somewhere else, even at my age.
>
> (Linda, 2017)

Linda's difficult experience of return resonates strongly with other interviewees in my wider sample, but especially some who also returned as single women in later life having lived for a long period in one country overseas. For these women, 'home' had been built among friends in their country of residence and fitting back into the life of a retiree in the UK was experienced as a challenge, adding further evidence that return is far from a straightforward process in terms of the unsettling experience of making the necessary personal and social adjustments (Percival 2013). For Linda, return migration is inconclusive. Hammerton and Thomson (2005, 312) note that a significant number of those returning from Australia chose to later re-emigrate, while many others live with a sense of regret and continue to 'journey [...] in their imagination'.

In particular, Linda's account revealed the difficulties of her experience of making a social life on return:

> In Dubai, or when you're living abroad, everybody is more friendly, because you don't have your family there. I was so disappointed in England: you go out in the rain, you get into the car, and you drive to this church hall, and you sing for two hours, and you try to talk to people and they're all dashing off home, and I invited some people to my house to coffee one afternoon or one evening and they looked at me like I was a lunatic! I tried several of these choirs and, sadly, I gave up, because I just couldn't deal with it. I found it very difficult to make new friends. I think when you're working and in your working environment, it's much easier to make new friends ... Even though I'm a retiree, I don't like retiree clubs in England because they feel a bit too old for me, I mean that sounds a bit silly I know.

Linda's continuation of friendships with other returnees was something that every interviewee, without exception so far, has commented upon as being important (see also Knowles 2008). Many of the people I interviewed had also found that any new friends they had made were also repatriates. They found residential clusters of other returnees on purpose or by accident, or sought out other repatriates through a special interest club (such as Bridge or mah-jong).

Patricia's return migration was also not conclusive, but for very different reasons. She and her husband continued a transnational living arrangement, travelling regularly to the UAE, especially during the UK winter. Nevertheless, fitting in back in the UK was still important to her and Patricia identified a sense of relief that she had found the *right* kind of older people to live among:

> I think this is the perfect place if you're going to retire as an outsider, there are quite a lot of newcomers in the village and there are quite a lot of old people, and it's a very interesting mix, they're nearly all retired professionals. There are a lot of retired teachers, an admiral, and doctors and dentists and things, and they're extremely nice. It's a very close-knit village. They're a very open-minded community; they're not people who have been here all their lives and not been out of Somerset. They're all very worldly people I think. I think we're lucky to be here, really, otherwise it would have made settling back more difficult.
>
> (Patricia, 2017)

As Caroline Knowles (2008, 177) suggests in relation to British colonial service returnees settling in rural England, return migrants attempt 'fitting in and being different' at the same time, efforts that rely upon the inhabitation of a classed and racialised social location while also marking ones' life as 'exotic'. Returning to a rural village on the south coast of Britain, Patricia's return echoes that of previous generations. A colonial elite materialised whiteness in the landscape through country houses funded by colonial wealth and displayed the artefacts of empire within them (Finn and Smith 2018; Harper 2005; Knowles 2008). Yet Patricia does not intentionally or strategically draw upon this resource. Furthermore, while Patricia feels welcomed by the professional retired community, she does not identify with them:

> It sounds strange but I always say, they have these village functions, they have tea parties and walks and treasure hunts, and social things, and I say to Robert, 'they're all *so* old,' but he said: 'hang on, they're our age!' [Laughs]. That is the big difference from being here and being in Dubai where our friends are all so much younger. There's no ageism there either: the people that are friendly with us they don't see us as being old fogeys.
>
> (Patricia, 2017)

For return migrants like Patricia, ageing is negotiated in relation to the cultural norms and ideologies of ageing among two social groups: the migrant community (since contact with Emirati nationals is minimal) and the society at home (Horn and Schweppe 2015). In her narrative, Patricia emphasises that *chronological* age does not determine her outlook or activities and her transnational living arrangement allows her to resist the stereotypes of older age she associates with later life in the UK by distinguishing her international social life.

Intriguingly, James instead perceived this same age structure in Dubai as a barrier to residence there:

> There aren't that many people who stay on there after retirement, so to find people in your age group. There's loads of things to do there, but it's mostly geared towards the younger people, when I say younger, I mean any age up to 50. I think I would feel out of place living in Dubai now. I probably wouldn't want to join anything that I used to be in because they are all 40 or 50.
>
> (James, 2017)

James, like other British male returnees who frame their life-stories through work mobilities, provides evidence of the central place that international career trajectories can occupy in people's identities. Upon return, the recounting of his migration contributes to a sense of achievement and a 'life well-lived' (Hammerton and Thomson 2005). He was therefore able to celebrate returning to the UK – 'If you've been a civil engineer and been standing out in the sun for years, it's nice not to have the sun' – and he mentioned the opportunities to enjoy cultural activities – 'classical music concerts, opera, theatre, museums, and art galleries' – that weren't available in Dubai at the time he returned. He chose to live near his late father and his sisters in Yorkshire, rather than returning to London (where he had owned property while working overseas) or relocating to the south coast of England. Like Linda, he also continues to socialise mainly with family and friends made in Dubai or other postings, more frequently with those who live near him but also travelling around the UK and overseas. However, as for others in my wider sample of returnees, experiencing illness in older age made James feel a stronger sense of loyalty to his residence in the UK and perhaps transformed the way in which he framed his account. The UK afforded him access to the NHS health and care services that he received when he discovered a heart problem shortly after retirement. Many retired lifestyle migrants continue to have second homes in the UK as a 'safety net' in older age and returning for state-funded care and healthcare support in their own language is a very common reason for return (Hall, Betty and Giner 2017).

Concluding remarks: the vulnerabilities of retirement, return and later life

Hall, Betty and Giner (2017) argue that the level of choice and control over return migration varies among Britons in later life, with some feeling that their return is involuntary. Their study in a European context emphasises the impact of ageing, especially associated health concerns: advanced older age often brings with it a demand for care that, in turn, prompts Britons to return to the UK to utilise state welfare provision and family networks. In contrast, the three Britons discussed in this chapter were relatively young and faced no such healthcare concerns prior to their return, yet ageing was still a significant factor shaping their return migration. Return decisions among these British migrants, were directly informed by transition to retirement age in the UAE, since the *Kafala* sponsorship system links residence directly to their employment through the issue of visas by the government. Individual resources and capitals (financial, occupational, and social) then informed how each individual could respond to and navigate this legislation, and the extent to which they could choose their exact age of retirement and return (see also Botterill 2016). Patricia's investor status as a business and property owner in Dubai cushioned her (and her husband) from experiencing the same kind of 'forced' return as Linda in later life. Indeed, Patricia was able to enact a transnational and seasonal living arrangement by returning to holiday in what became her second home in Dubai. James fell somewhere between Linda and Patricia in terms of his sense of agency over returning: while the economy was buoyant, his skills as a project manager were in high demand and his contracts renewed in spite of being over the formal retirement age. Later, in the post-economic crisis period, he experienced how his labour was expendable and his age a barrier to continued employment in the UAE, but he was able to reorient his career trajectory elsewhere one final time.

None of the British migrants I have interviewed at all in the wider study were able to fully choose the terms of their residence or return, but it is still striking that all three accounts from migrants to the UAE suggest such reluctance. Examining the return journeys made by migrants, reminds us that *voluntary* economic migration is never a singular decision made once but instead shares in common with lifestyle migration an on-going comparative evaluation of where life is (imagined to be) 'better' (Benson and O'Reilly 2009). Yet, while a sense of choice saturates the analytical framing of British migration of this kind, the global economy and legislation of particular states set parameters in which such choices emerge. Certainly, in everyday day life in the UAE, ethnicity, nationality and racial markers have been demonstrated to make the white 'expatriate' experience markedly different of other migrants, even those with similar professions and income (Vora 2013; Walsh 2018). Yet taking age into account nuances this assertion. British subjectivities in the Gulf are also marked by a degree of vulnerability when compared with citizens who have fuller social rights. This vulnerability is revealed when they reach retirement age and may continue after return. The narrative accounts demonstrated that British returnees sometimes find it difficult to 'fit in' and socialise with others who never left Britain, such that their sense of belonging in the UK is brought into question. Therefore, in order to effectively elucidate the complex social locations of British migrants, we will need to pay attention to processes of ageing and their intersections. Nuancing accounts of the privilege of British migrant subjectivities through a recognition of the vulnerabilities of later life, raises new questions about the relationship between home, settlement and citizenship.

Note

1 Patricia, James and Linda did not have the financial resources or social networks of an established or elite British middle class. They were not among the highly mobile and extremely high-income employees of global corporations, such as the finance workers who currently dominate the globalisation literature (Beaverstock 2005). They were not privately educated, did not have access to family wealth, and were also not connected with the diplomatic service. Instead, they were among a generation who experienced significant social mobility, first through their education (in John and Patricia's case a grammar school education that led to university) and then through their international migration. Their three interview accounts cannot be representative of the full internal diversity of the middle-class British community in Dubai. Consequently, although I make some observations about the impact of occupation and property ownership on British migrants' capacity to negotiate their visa status in the UAE, the aim in this article is not to systematically examine how class shapes return.

References

Ali, S. 2010. *Dubai: Gilded Cage.* London: Yale University Press.

Beaverstock, J. 2005. 'Transnational elites in the city: British highly-skilled inter-company transferees in New York city's financial district'. *Journal of Ethnic and Migration Studies*, 31(2), 245–268.

Benson, M. and O'Reilly, K. eds. 2009. *Lifestyle Migration. Expectations, Aspirations and Experiences.* London and New York: Routledge.

Botterill, K. 2016. 'Discordant lifestyle mobilities in East Asia: privilege and precarity of British retirement in Thailand'. *Population, Space and Place*, 23(5).

Buckley, M. 2012. 'From Kerala to Dubai and back again: migrant construction workers and the global economic crisis'. *Geoforum*, 43(2): 250–259.

Buffel, T. 2015. 'Ageing migrants and the creation of home: mobility and the maintenance of transnational ties'. *Population, Space and Place*.

Coles, A. and Walsh, K. 2010. 'From "trucial state" to "postcolonial" city? The imaginative geographies of British expatriates in Dubai'. *Journal of Ethnic and Migration Studies*, 36(8): 1317–1333.

Conradson, D. and Latham, A. 2005. 'Transnational urbanism: attending to everyday practices and mobilities'. *Journal of Ethnic and Migration Studies*, 31: 227–233. doi:10.1080/1369183042000339891.

Conway, D. and Leonard, P. 2014. *Migration, Space and Transnational Identities: The British in South Africa.* Basingstoke: Palgrave Macmillan.

Davidson, C. 2005. *The United Arab Emirates. A Study in Survival.* Boulder CO: Lynne Rienner Publishers.

Fechter, A. and Walsh, K. 2010. 'Introduction to Special Issue: examining 'expatriate' continuities: postcolonial approaches to mobile professionals'. *Journal of Ethnic and Migration Studies*, 36(8): 1197–1210.

Finch, T., Andrew, H. and Latorre, M. 2010. *Global Brit: Making the Most of the British Diaspora.* UK: Institute of Public Policy Research.

Findlay, A. 1988. 'From settlers to skilled transients: the changing structure of British international migration'. *Geoforum*, 19(4): 401–410.

Finn, M. and Smith, K. eds. 2018. *The East India Company at Home 1757–1857.* London: UCL Press.

Forrest, R. and Hirayama, Y. 2015. 'The financialisation of the social project: Embedded liberalism, neoliberalism and home ownership'. *Urban Studies*, 52(2): 233–244.

Giner-Montfort, J., Hall, K. and Betty, C. 2016. 'Back to Brit: retired British migrants returning from Spain'. *Journal of Ethnic and Migration Studies*, 42(5): 797–815.

Hall, K. and Hardill, I. 2014. 'Retirement migration, the "other" story: caring for frail elderly British citizens in Spain'. *Ageing and Society*, 36: 562–585.

Hall, K., Betty, C. and Giner, J. 2017. 'To stay or to go? The motivations and experiences of older British returnees from Spain'. In *Return Migration and Psychosocial Wellbeing*, edited by Z. Vathi and R. King, 221–239. London: Routledge.

Hammerton, A. and Thomson, A. 2005. *Ten Pound Poms. Australia's Invisible Migrants.* Manchester and New York, NY: Manchester University Press.

Harper, M. ed. 2005. *Emigrant Homecomings: The Return Movement of Emigrants, 1600–2000.* Manchester and New York: Manchester University Press.

Harvey, W. 2009. 'British and Indian scientists in Boston considering returning to their home countries'. *Population, Space and Place*, 15: 493–508.

Heard-Bey, F.1982. *From Trucial States to United Arab Emirates: A Society in Transition*. London: Longman.

Hollway, W. and Jefferson, T. 2000. *Doing Qualitative Research Differently: Free Association, Narrative and the Interview Method*. London: Sage.

Hopkins, P. and Pain, R. 2007. 'Geographies of age: thinking relationally'. *Area*, 39(3): 287–294.

Horn, V. and Schweppe, C. eds. 2015. *Transnational Aging: Current Insights and Future Challenges*. New York, NY: Routledge.

King, R., Warnes, T. and Williams, A. 2000. *Sunset Lives: British Retirement Migration to the Mediterranean*. Oxford: Berg.

King, R., Lulle, A., Sampaio, D. and Vullnetari, J. 2017. 'Unpacking the ageing–migration nexus and challenging the vulnerability trope'. *Journal of Ethnic and Migration Studies*, 43(2): 182–198.

Knowles, C. 2005. 'Making Whiteness: British lifestyle migrants in Hong Kong'. In *Making Race Matter: Bodies, Space and Identity*, edited by C. Knowles and C. Alexander. London: Palgrave Macmillan.

Knowles, C. 2008. 'The landscape of post-imperial whiteness in rural Britain'. *Ethnic and Racial Studies*, 31(1): 167–184.

Knowles, C. and Harper, D. 2009. *Hone Kong: Migrant Lives, Landscapes, and Journeys*. Chicago andLondon: Chicago University Press.

Leonard, P. 2010. *Expatriate Identities in Postcolonial Organizations: Working Whiteness*. Aldershot, UK: Ashgate.

Mohammad, R. and Sidaway, J. 2016. 'Shards and stages: migrant lives, power, and space viewed from Doha, Qatar'. *Annals of the American Association of Geographers*, 106(6): 1397–1417.

Nare, L., Walsh, K. and Baldassar, L. 2017. 'Ageing in transnational contexts: transforming everyday practices and identities in later life'. *Identities: Global studies in culture and power*, 24(5): 515–523.

O'Reilly, K. 2000. *The British on the Costa Del Sol: Transnational Identities and Local Communities*. London and New York: Routledge.

Oliver, C. 2008. *Retirement Migration: Paradoxes of Ageing*. London: Routledge.

Percival, J. 2013. '"We belong to the land": British immigrants in Australia contemplating or realising their return "home" in later life'. In *Return Migration in Later Life*, edited by John Percival, 113–139. Bristol: Policy Press.

Reuters 2016. *Could you retire in the UAE?*27 September 2016. https://mena.thomson reuters.com/en/articles/retire-in-uae.html.

Rogaly, B. and Taylor, B. 2009. *Moving Histories of Class and Community. Identity, Place and Belonging in Contemporary England*. Basingstoke: Palgrave Macmillan.

Saunders, P. 1989. 'The meaning of "home" in contemporary English culture'. *Housing Studies*, 4: 177–192.

Vora, N. 2013. *Impossible Citizens: Dubai's Indian Diaspora*. London: Duke University Press.

Walsh, K. 2016. 'Expatriate belongings: traces of lives "abroad" in the home making of English returnees'. In: *Transnational Migration and Home in Older Age*, edited by K. Walsh and L. Näir, 139–151.

Walsh, K. 2018. *Transnational Geographies of the Heart: Intimate Subjectivities in the Globalising City*. Oxford: Wiley Blackwell.

Warnes, A., King, R., Williams, A. and PattersonG. 1999. 'The well being of British expatriate retirees in Southern Europe'. *Ageing and Society*, 19(6): 717–740.

Index

Note: Information in tables is indicated by page numbers in **bold**.

Fat Studies 147
Fechter, Meike 63
feminist scholarship 28, 148, 150, 155
Findlay, Allan 4
flags 135–136, 137–138
France 24–35

gender: difference and 159; embodiment
   and 148; fatness and 149; health and
   150; intersectionality of 158; processes
   of 146; social mobility and 27
globalisation 33
'good migrant' 34

homeownership 48–52, 189–191
Hong Kong 84

imperialism 11, 96, 131–132; *see also*
   colonialism
India 115–125, **118**
integration: into local community 28,
   32–35
International Student Mobility (ISM) 7
interracial marriage 97–99, 112n2
intra-European migration 29–30
ISM *see* International Student
   Mobility (ISM)

Kenya 75–90
Knowles, Caroline 194

landscape 31–32
Larsen, Jonas 62
legal liminality 85–88
lifestyle migrants 4–5, 25–26, 44–45, 78,
   96; brunch and 62–63; and 'ladies who
   lunch' 63–64
lifestyles: of expatriates 61–67
liminality: legal 85–88; postcolonial
   88–90
local community: integration into 28,
   32–35
Lot 24–35

Malan, Daniel 98–99
marriage: interracial 97–99, 112n2;
   women and 15
Mayle, Peter 35
middle class 27–29, 34
middling transnationals 5–6
migration: body size and 148–149;
   consumption led 43–45; 'good' 34;
   intra-European 29–30; lifestyle 4–5,
   25–26, 44–45, 78, 96; retirement 5, 78,

117, 124–125; return 7, 182–196;
   reverse 6; as right 32; settler 3–4;
   vulnerability and 14–17

'narrative of decline' 135
New Zealand 128–140
normativity, transnational 94–112
nostalgia 36
nurses 41

obesity 146–160
obesity epidemic 146–147
OCI *see* Overseas Citizen of India (OCI)
   card
older age 77–78, 164–167, 164–179,
   168–171, 182–196, 184–185, 188–192,
   195–196; *see also* retirement migrant
ontological security 191
O'Reilly, Karen 25
Other 10, 30, 59, 83, 155–156
Overseas Citizen of India (OCI) card 116

Pākehā culture 136, 141n6
Panama 25
parents' expectations 120–121
patriotism 13, 129, 131, 134–136,
   137–138
Permanent Residency 69
Perth, Australia 40–54
political journeys 106–110
postcoloniality 30–31, 36, 78, 83–85
postcolonial liminality 88–90
practices: defined 59; expatriates and 81;
   local community and 60; *see also*
   lifestyle migrants
privilege: colonialism and 36, 76–77;
   expatriates and 8–10; quality of life
   and 25–26; whiteness and 97–99
professional migrants 4, 183
public sector employment 27

racialisation 29–30, 34–36, 87–88, 99,
   156–157, 159–160
racial journeys 106–110
racial solidarity 76
resilience 164–179; community 175–177;
   family 171; individual 168–170; vulnerability and 168–171
retirement 182–196
retirement migrant 5, 78, 117, 124–125,
   165; *see also* older age
return migrants 7, 182–196
reverse migration 6
right: migration as 32

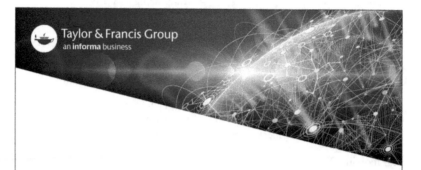